GRAPHICS AND SPACE

GINGKO PRESS

Graphics and Space

ISBN: 978-1-58423-389-3

First Published in the United States of America by Gingko Press
by arrangement with Sandu Publishing Co., Limited

Gingko Press, Inc.
1321 Fifth Street
Berkeley, CA 94710 USA
Tel: (510) 898 1195
Fax: (510) 898 1196
Email: books@gingkopress.com
www.gingkopress.com

Sponsored by: Design 360°– Concept and Design Magazine
Chief Editor: Wang Shaoqiang
Executive Editor: Daniela Huang
Chief Designer: Angel
Book Designer: Antiny Wu
Sales Managers:
Niu Guanghui (China), Daniela Huang (International)
Address:
3rd Floor, West Tower,
No. 10 Ligang Road, Haizhu District,
510280, Guangzhou, China
Tel: (86)-20-84344460
Fax: (86)-20-84344460
sandu84344460@163.com
www.sandu360.com

Cover project by Tobias Rehberger and Artek

Printed and bound in China

Contents

Sebastien Agneessens
Formavision, USA

Sebastien Agneessens is the creative director & president of Formavision, a New York-based creative agency that collaborates with artists and designers to develop and produce innovative artistic projects that capture the imagination, reinforce the brand vision and resonate on an emotional level. Formavision is uniquely positioned to understand the dynamics of both the corporate and the artistic spheres, fostering synergies between business strategies and artistic visions. All Formavision projects invoke creativity, authenticity and self-expression, creating a cultural infrastructure within which people experience the brand's identity and aspirations.

Preface

The wake of the new millennium will be remembered in large part for the rising power of corporations, for the monumental scale of international architectural projects, the boom of the global art market, and the acceptance of design as the common language of consumerism and good taste in post-industrial societies. Facing increasingly educated, sophisticated and aware markets, brands are challenged to push their boundaries and integrate the vocabulary of design in all of their expressions, including retail. The nomenclature "retail design" is gradually being replaced by "experiential design," a subtle switch in the practice that infuses the retail experience with flavors of music, art, lifestyle, illustration and design, reflecting a certain sense of involvement in culture and entertainment.

While lifestyle brands march into this promising retail territory, designers and creative directors are encouraged to experiment in the way they approach their work by broadening the array of their practices. I am personally part of this generation of creatives who have been commissioned to develop retail projects even though I was never trained as an architect or an interior designer. What I had instead was a growing following of my studio's work coupled with a desire to expand my playground into the various arenas of what has now become experiential design: sound, smell, graphic design, illustration, industrial design, craft. I use myself as an example because several of the creatives featured in this book share a similar background, fueled by a desire to express themselves through all media and expand the realm of their practice. This inter-disciplinary approach, a sort of twenty-first century renaissance, if you will, is what brands are looking for today in order to maximize the impact of any of their retail projects.

It is important to comprehend this trend not only through the filter of pure design, but also through a wider cultural lens. The approach of design for the sake of mere design is indeed a superficial and short-lived argument, and brands needs to delve into cultural, sociological or artistic references in order to resonate in the minds of their target audiences. Since brands have for better or for worse become natural extensions of people's expressions and identities, they need to align themselves culturally with their stakeholders in all of their communication efforts, and in particular through their retail applications like the ones featured in this book. Oddly enough, even non-brands such as Maison Martin Margiela need to signify their opposition to the fundamentals of branding in order to be accepted by their anti-branding peers.

In the same way that brands have been fine-tuning their identities, designers have also been able to assemble their own languages by creating a vocabulary of illustrations, objects, spaces, colors, shapes and tones. Some renowned designers like Jaime Hayon, Marcel Wanders or Karim Rashid have even become their own brands, and are hired by corporate clients not only because of their styles, but also because of their following, their press value and perceived personalities. This phenomenon is similar to co-branding in a certain sense, as the importance of the designer is as valuable as the prestige of the brand, and the projects are about mélange of both worlds in the quest for freshness, fun, newness and good design.

In the search for self-expression, audience engagement, press and street attention, retailers have pushed the graphic elements of architectural design to new heights, as visual information is applied to every surface possible in order to maximize visual impact. Now that we can print on most substrates, from wood, vinyl, fabric, and glass to metal, printed presence can be increased to the maximum. The ubiquity of graphic art is not only enabled by technology, but also by the cultural phenomenon of Street Art, which though initially marginalized has now become widely popular and accepted by both the establishment and the art world, making desirable the complete covering of surfaces with illustrations. What was once a symbol of rebellion and vandalism is now an artistic gesture.

Let's remember that the act of drawing goes back to the roots of our humanity, constituting a trace of our ephemeral existence and symbolizing our passage on earth. With the drawings of the Lascaux caves in mind, or the Mayan or Egyptian hieroglyphs, the frescoes of Pompei, the Buddhist murals of Pagan, the marble floor of the Sienna cathedral and the adorned ceilings of the New York Deco hallways, drawings have always figuratively or abstractly represented the beliefs and states of mind of the periods in which they were conceived. From the selections featured in "Graphics and Space," I see on the one end an expression of systemic interconnectivity, a probable result of the Internet, information technology and more generally globalization, and on the other end a wishful return to simplicity, to nature, a hope for balance between people and their environment. How both these movements will coincide is the adventure we are facing in the twenty-first century.

Design Agency | Projekttriangle Design Studio **Creative Direction |** Martin Grothmaak **Design |** Franz Stämmele **Photography |** Projekttriangle Design Studio, Breil+Partner **Client |** Son's & Daughter's **Location |** Kehl, Germany

Son's & Daughter's In-store Graphics

Son's & Daughter's is a German label for children's fashion and accessories. Breil+Partner developed an unconventional interior concept for the store, which comes without common childish standard patterns and thereby appears in contrast to other solutions in this area. Projekttriangle Design Studio was commissioned to design a graphical in-store campaign for the label; they created more than 50 characters with thematic worlds around them. The unusual looking creatures are fixed as film plots in black, orange and white onto walls, changing rooms and furniture. Assimilated to their built environment in an amusing manner, they fly from the ceiling, climb in shelves and loll around the corners. Besides their funny appearance, the little creatures guide visitors through the entire shop, referring to its special highlights as well as to functional areas; for example, the kid's play area and the multimedia station. Because of their flexibility in production and design, they can be adapted to changing situations on-site. New stories can be generated over and over again. In addition, the characters are also a recognizable feature on merchandise products and print applications.

CLOTHING

Design | ZEK Crew **Photography |** Jaka Vinšek
Client | 3FS **Location |** Kranj, Slovenia

3FS

Graphic design of 3FS (a mobile/web/game development company). The office walls are covered with custom cut black and white stickers.

Esc

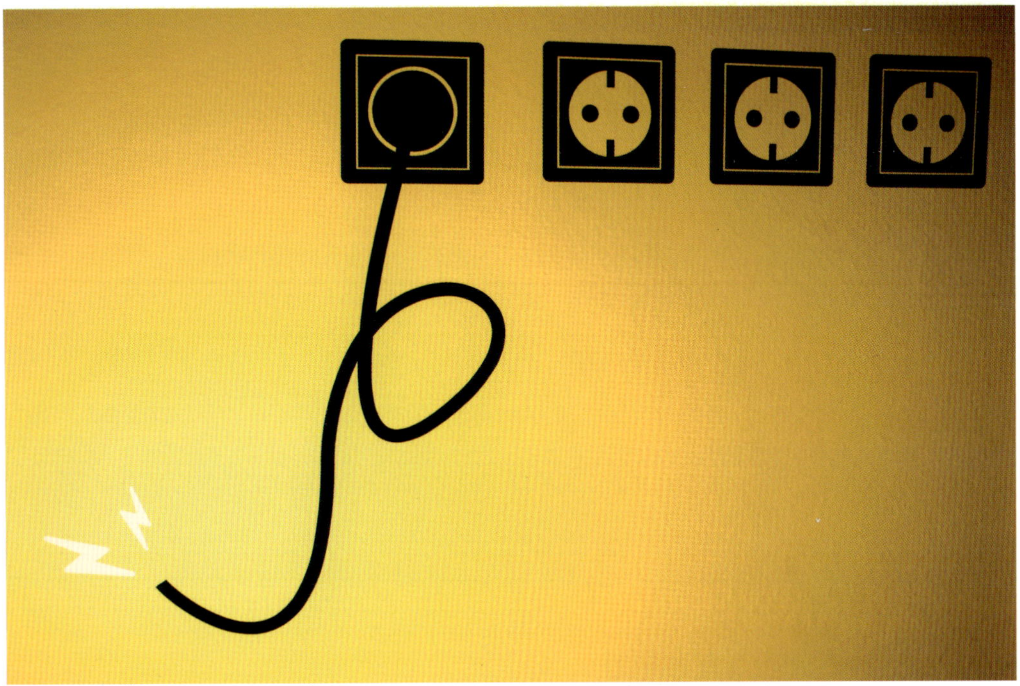

Design | Branimir Sabljic **Client |** Advertising agency "Pokobar" **Location |** Zagreb, Croatia

Pokobar Interior Design

In the downtown of the city of Zagreb, an apartment of 180 square meters was turned into an office space for three sister companies. The designer decided to express the typical division of rooms according to function by spelling out each room's function on the wall in hypertrophied typography. The effect of this "indoor graffiti" transforms the segmented apartment into a public space where business is conducted, while retaining a frivolous style used by creative professionals. Each room is styled differently to show the function, with a combination of hand painting and disrupted straight wall lines and angles setting the mood.

office
office

ENERGETICS

BOK

Design | Branimir Sabljic **Photography** | Damir Fabijanic, Branimir Sabljic **Client** | Studio Rasic **Location** | Zagreb, Croatia

Art Of Persuading Exhibition

Creation of an entrance room for the historically significant "Umjetnost Uvjeravanja" (Art of Persuading) exhibition which celebrates 170 years of advertising in Croatia. The "Graffiti Room" is the introduction to the exhibition. The space graphically illustrates the five sections: 1) target groups; 2) method/communication channels; 3) messages/meanings; 4) media and 5) advertising politics and policy.

MEDIJI
medium
OGLAŠAVANJE
POLITIKE I
POLITIKA
OGLAŠAVANJA
advertising politics
and the politics of advertising

CILJANE SK
target groups

MEDIJI
medium

UVJERAVANJA
Oglas

CILJANE SKUPINE
target groups

Creative Direction | Mattias Mohr **Design |** ZMIK **Photography |** Eik Frenzel
Client | Sevensisters **Location |** Basel, Switerland

Art Shop 09

During the Art|40|Basel (Switzerland) ZMIK was called on to create an appealing art shop in a cloakroom which is separated from the main visitor circulation route by a long, wide staircase. The design aims to attract customers from the ground floor and to transform an unalluring cloakroom into a temporary design boutique.

Hundreds of black arrows draw the visitor's attention and point the way upstairs to the art shop. The self-adhesive arrows cover walls, floor and ceiling. Towards the center of the shop they condense and climax around the closely-packed display. The Batman & Robin lamps hover in a dense arrangement above the three display volumes and highlight the colorful products.

Base
Art|40
ART BAG

Art
Bas
Art | 40
Base
Base
Base

Basel
Art|4
Basel

Design | Momorobo **Illustration** | Sally Zou, Morris Lee **Client** | Nike
Location | various cities in China

Nike: Back to School

The brief was to create a single key visual for a retail campaign, Back to School, to promote Nike products (e.g. bags and shoes) to students during the summer holidays. The key visual was to be utilized on an installation made up of desks, to be placed and exhibited at select Nike stores in more than 10 cities in China, and also as a canvas to display the energy and lifestyle of Chinese campus life. A wooden pencil case was also produced with the visuals silk-screened on it, as a giveaway for consumers who purchased the products.

The illustration is a construction of 3 different parts, which Momorobo felt is the perfect representation of Chinese campus life: going to school, classroom lessons and physical education. It is also a fusion of eastern mythology figures such as the Dragon, Qilin etc, and Chinese pop culture films like *Palm of Ru Lai* (a Hong Kong Wuxia film released in the 1960s), and *Shaolin Soccer* (a movie directed by and starring Stephen Chow). Kung Fu was also subtly integrated with sports moves to enhance the entertainment value of the illustration.

A treatment that looks like a drawing from a marker is a cheeky reference to restless students who draw on their classroom desks.

yo!

PACK
YOUR
LIFE.

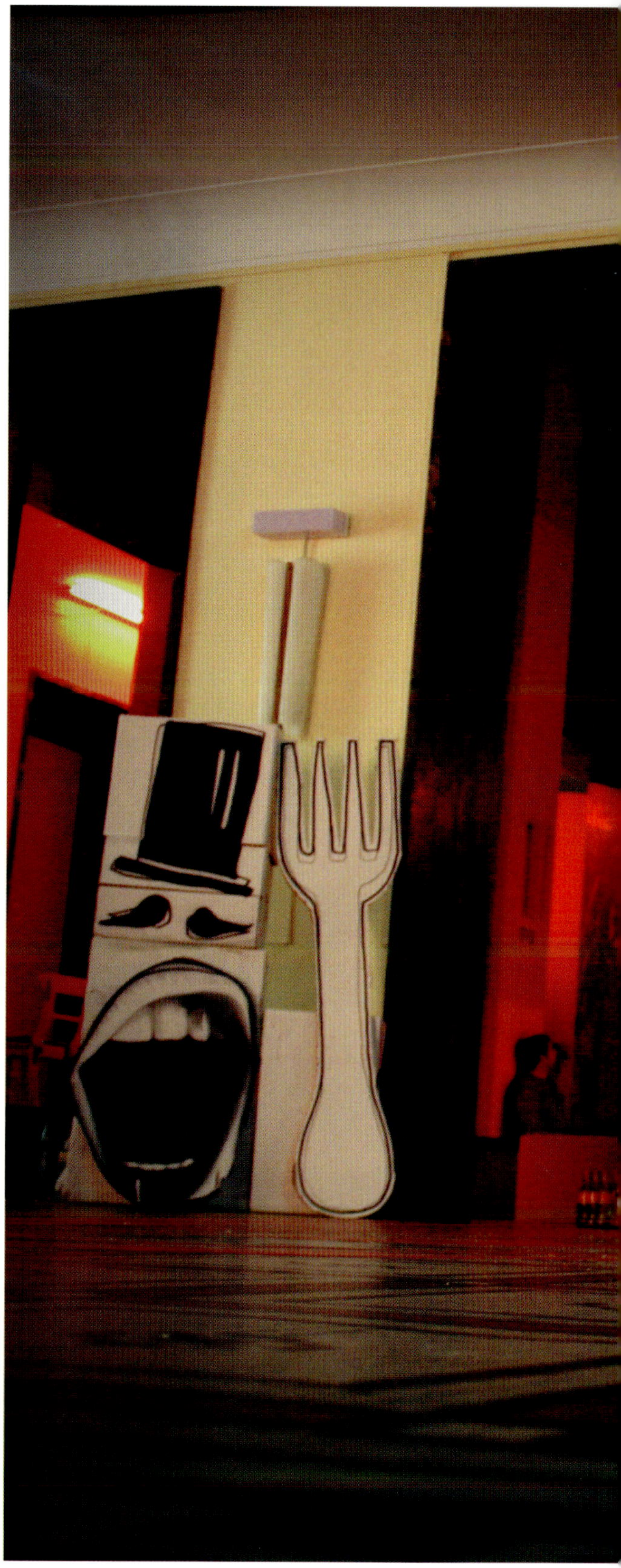

Design | Janina Sitzmann, David Lüpschen **Location |** Cologne, Germany

SoCo Limelight Tour

Design for SoCo Limelight Exhibition.

Design | David Lüpschen **Location |** Cologne, Germany

Levi's Blue Lab Store

Fashion Meets Art – The Levi's® Blue Lab creates the store design for a limited period of three months. It is a platform for fashion and art.

3

WWW.JIAR.DE
JIAR
GRAFI
KFITI

Design | David Lüpschen **Location |** Brussels, Belgium

No New Enemies "A Little Scratch X-mas"

Wall installation and window display design for No New Enemies "A Little Scratch X-mas."

mr. ego
mr. ego
mr. ego

JIAR
grafikfiti*

mr.
ego
mr.
ego
mr.
ego

Design | Good Wives and Warriors **Photography |** Venetia Van Hoorn **Client |** Space In Between Gallery **Location |** London, UK

Buckminsterfullerene Dream

Buckminsterfullerene Dream was a pop-up exhibition in conjunction with 'Space In Between Gallery,' in an unused shop space on Clerkenwell Road, London. It consisted of a floor painting, a 3D drawing sculpture, and a wall painting created directly on all of the walls and columns of the space. The work was a continuation of Good Wives and Warriors' work based on the 'origin of life' theories.

Design | Via Grafik **Location |** Heilbronn, Germany

Burnside

When Via Grafik were asked to paint some walls in Heilbronns' skateshop, they tried to give consumers some orientation by referring to the function of the special sections of the shop.

Illustration | Miss Led **Location** | London, UK

D & Me Boutique

Miss Led had only two weeks to complete artwork on as much wall space as possible over 3 floors of adjoining staircases, corridors, doors and changing rooms. The works were produced from sketches created before and during the project. As the boutique was a unique feminine mix of new and vintage clothing and accessories, she really wanted to portray that in her illustration work, using influences from various times and places with a contemporary delivery. Materials used were aerosol paint and acrylic with brush, paint pens and chalk.

don't compromise yourself you're all you've got
GLAM
IGGY
MICK ROCK

You
Aint
Seen
Nothing

Wonderous
Delights
Still
In Sight

Design | thisismybworld **Location |** Athens, Greece

Baby Grand Hotel

A series of 7 hotel luxury suites painted in collaboration with the artist Zoe Zillion and curated by Breeder Gallery.

ZOE

Design | thisismybworld **Location |** Athens, Greece

Mezzo Art

An art cafe restaurant in the center of Athens, Kolonaki, painted in collaboration with the artist Stelios Faitakis and curated by Breeder Gallery.

Design | Steve Alexander (Rinzen) **Client |** Pictoplasma **Location |** Berlin, Germany

Plasmagoria

Rinzen were commissioned to design and paint the arena at Pictoplasma's Pictopia Exhibition at the Haus der Kulturen der Welt. Five bumper cars came to life for 10 minutes every 30 minutes throughout the course of the exhibition.

Creative Direction | Lesha Galkin **Design |** Dopludó Collective **Illustration |** Lesha Galkin, Egor Bashakov, Karina Eybatova **Photography |** Egor Bashakov **Client |** D Club **Location |** St. Petersburg, Russia

D Club & Café

Full space design of the small cozy club and café on the banks of the gulf of Finland in St. Petersburg. Elements include semicircular shapes, smooth transitions instead of a corner between the walls and ceiling and black murals on white walls, ceilings, and columns that depict images from magical fairy tales, northern folk stories, and mystical creatures.

Dalmatin
club

Design | Fabian Bertschinger & Manuel Schibli **Client |** Lucerne University of Applied Sciences and Arts
Location | Lucerne, Switzerland

Reception / Bar für Elia University Congress

The idea was to build a room which represents the Lucerne University of Applied Sciences and Arts, which is one of the leading art schools in Switzerland. It stands for a mix of analog and digital technologies. The classic paint and drawing training is integrated with modern computer technology. The designers developed a concept for the official bar and reception room of the Elia University Congress. A huge hand-drawn image decorates the walls, furniture, floor, and ceiling. The source is a digital printer hanging on the wall. The whole room and all of the furniture are painted completely white. Out of the printer comes a figure of lines, animals, clouds, explosions and plants. It stands for the interface between the digital and the analog world. A hidden spotlight with slowly changing colour transforms the atmosphere of the room and creates varying moods.

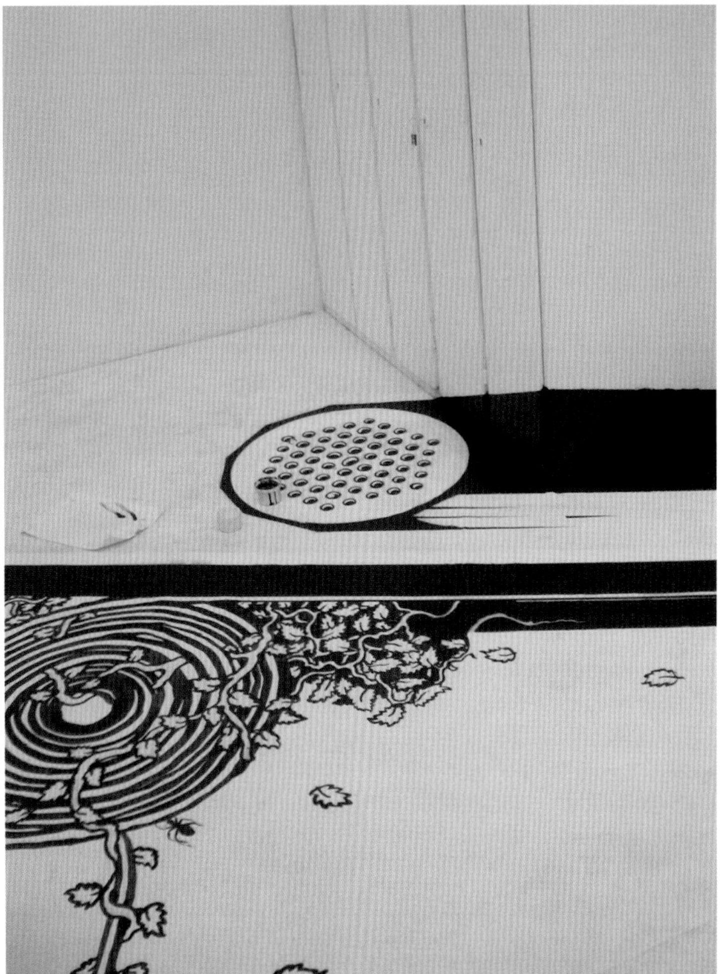

Illustration, Photography | Denise Fort **Location |** Munich, Germany

Design Parcours Exhibition

Design Parcours is an annual event which takes place in the center of Munich, Germany. A number of young designers, artists and agencies show their projects for ten days. Denise Fort drew a variety of images to create a strong visual experience.

UV1 F14

com

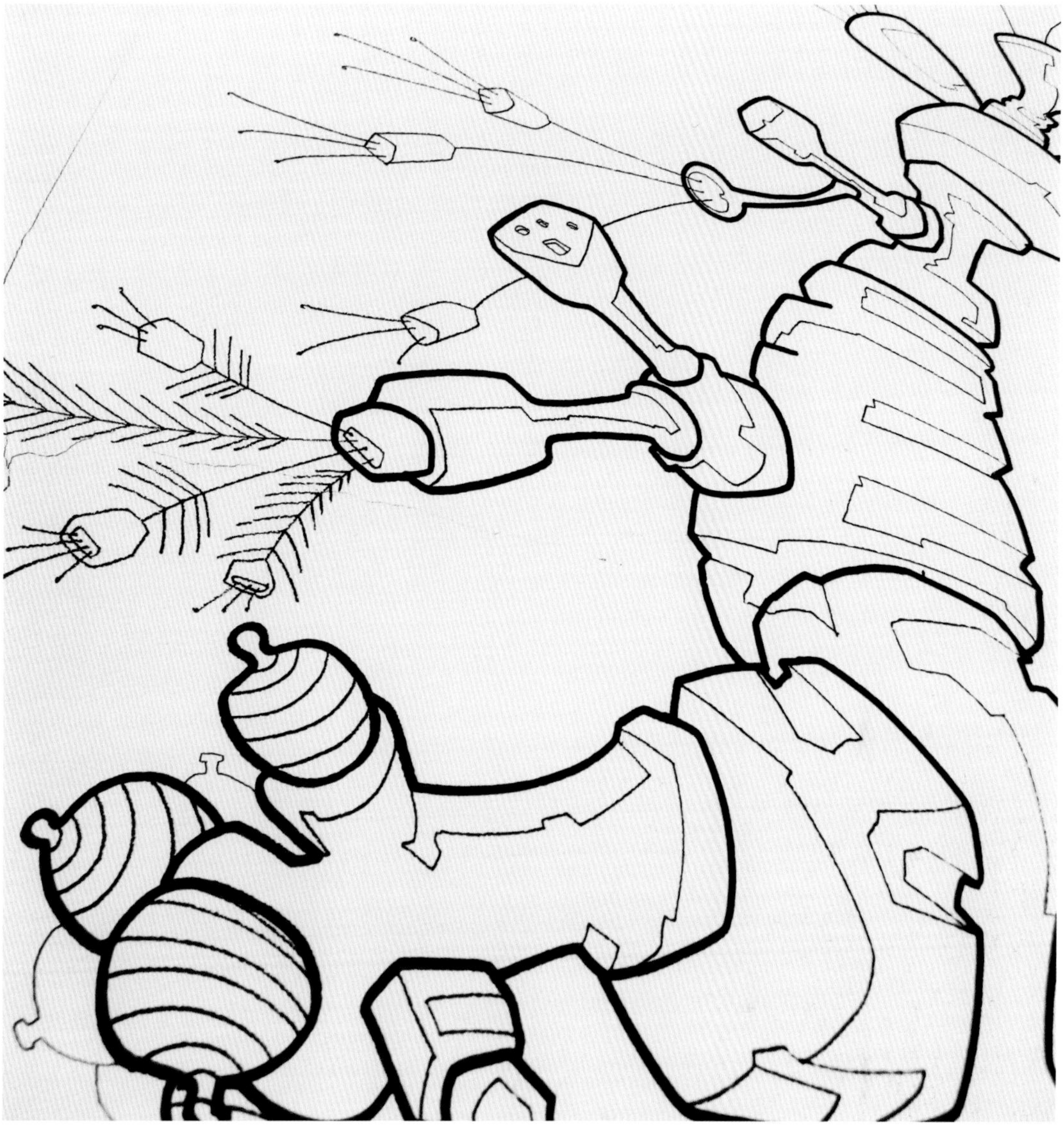

UV1 F13

Illustration | Lovisa Burfitt **Client |** H&M **Location |** Milano, Copenhagen, Tokyo and Seoul

H&M Graphic Communication for Flagship Stores

Lovisa Burfitt was asked to draw 50 illustrations around different themes for a selection of the H&M flagship stores around the world. The first opened was Milano (Italy), followed by Copenhagen (Denmark), Tokyo (Japan) and Seoul (Korea).

Lovisa Burfitt has won several awards for her H&M illustrations.

ACCESSORI

DONNA
UOMO
CABINE prova
DIVIDED
INTIMO!!!
CASSA

Illustration | Lovisa Burfitt **Location |** Stockholm, Sweden

East

In 2008-2009, Lovisa's artwork was exhibited on the walls of the restaurant East in Stockholm. Lovisa's illustrations were also shown on TV monitors.

mimi love sky

Let's go eskimo!!

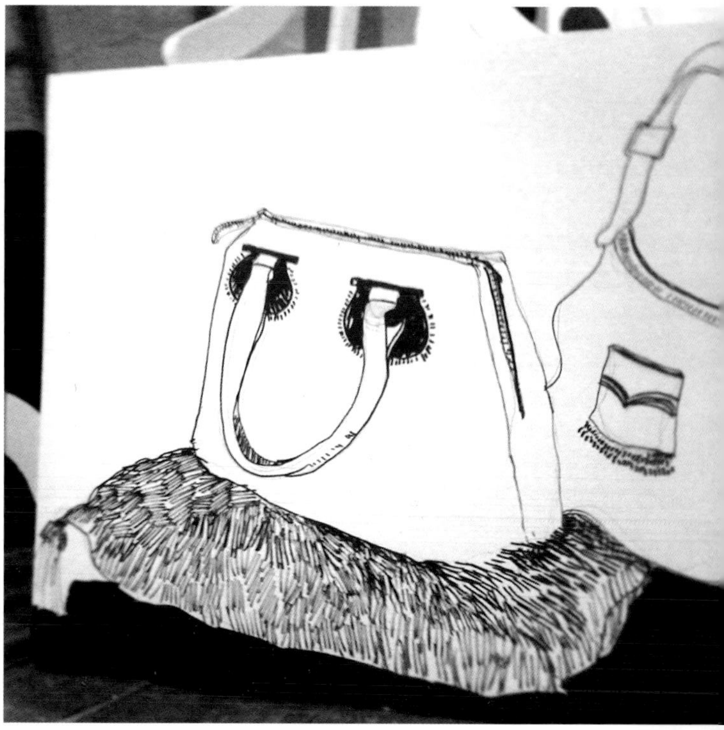

Design, Illustration, Photography | Camille Rousseau
Location | London, UK

Diesel Catwalk

Illustration for Diesel Catwalk.

Design | Mike Ballard **Location |** London, UK

The All of Everything

Mike Ballard was invited to do a solo show of his work at the Arts Gallery in University of the Arts in London. He decided to cover the entire gallery in one big painting and to cover the ceiling with a large digital collage. The piece is about transit space, and the temporality of life. Mike wanted to mummify the gallery space and prepare the walls for death. It signified the end of the beginning and the beginning of the end. Inside the Egyptian death mask, two videos and a soundtrack are played, providing a narrative for the show.

The building is now empty and awaits the demolition of the gallery. The exhibition ran for three months and received great reviews in the art press and from the visiting public.

THE END
OF THE
THE BEGINNING

Design Agency | Projekttriangle Design Studio **Creative Direction |** Martin Grothmaak **Design |** Danijela Djokic, Jürgen Späth, Franz Stämmele **Photography |** Thomas Mardo **Client |** Hugo Boss AG **Location |** Mannheim, Germany

Hugo Boss Orange Flagship Store

Projekttriangle Design Studio was hired for the interior graphic design of the new Orange Monostore in Mannheim, Germany. The entire project, of which Mannheim is a part, consisted of the development of a global, unified store campaign for the label Hugo Boss Orange. What they designed is a modular, graphical in-store concept, which has been implemented since 2006 in over 200 stores worldwide. In Mannheim, an over 40 foot high wall installation – the fantasy tree – consisting of 14 silk-screen printed panels and 24 broadcasting monitors, has been implemented. The installation acts as the central point of the 3-story, 1,300 square-foot shop. Detailed handmade illustrations from collages of fantastic figures and animals were applied onto the interior walls of the store and onto the fitting rooms and the store's furniture using different techniques. The use of digital media is also part of the concept. Depending on the shop's architecture, location, or cultural environment, illustrations are adapted through theme (such as a floral or faunal theme) or through variations in size, quantity, color, and materials.

BOSS
HUGO BOSS

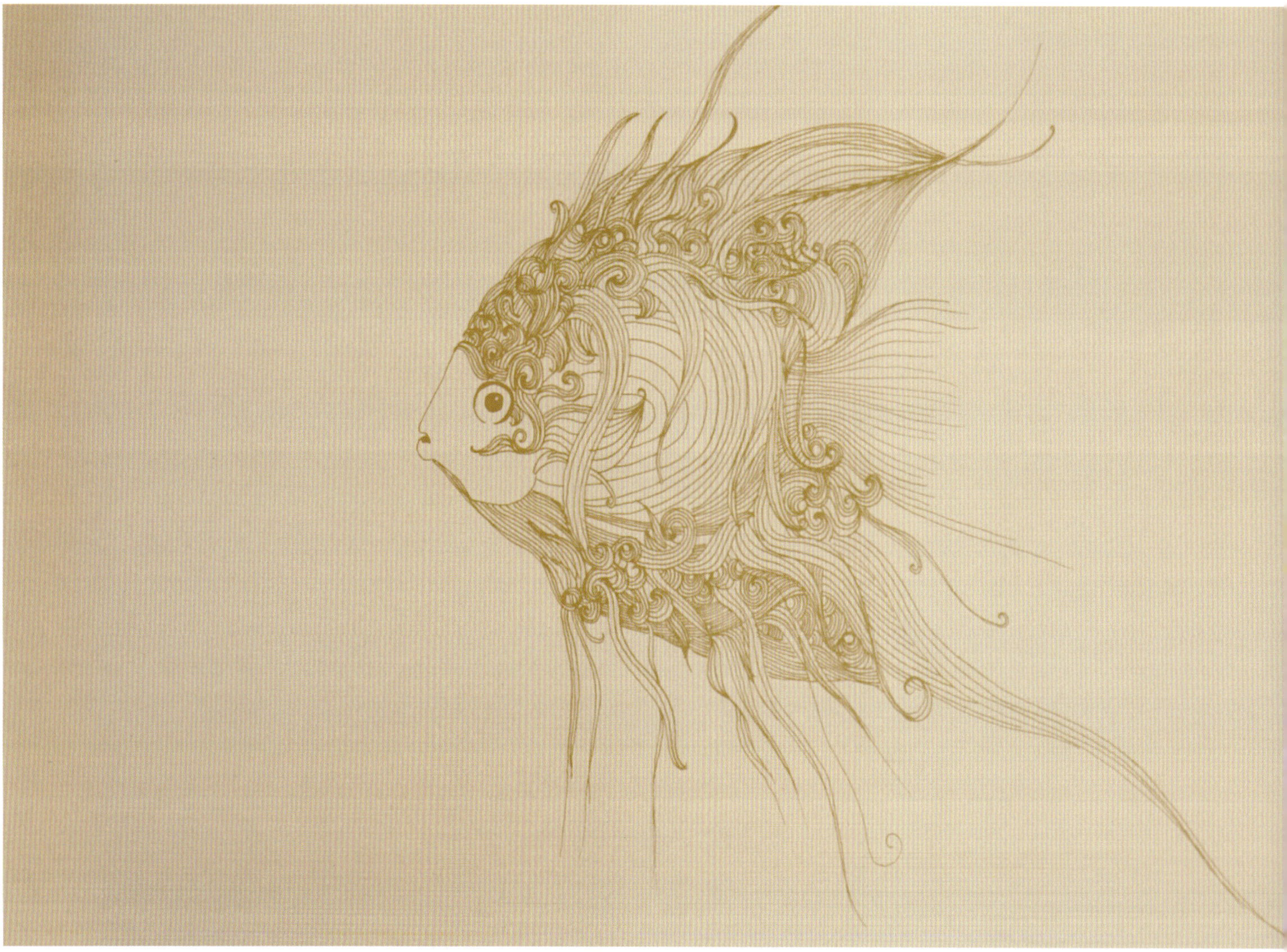

VORNE

Design Agency | Projekttriangle Design Studio **Creative Direction |** Martin Grothmaak **Design |** Danijela Djokic, Jürgen Späth, Franz Stämmele **Photography |** Tom Ziora **Client |** Hugo Boss AG **Location |** Shanghai, China

Hugo Boss Orange Special Concept Store

The Hugo Boss Orange Special Concept Store in Shanghai is part of the worldwide in-store campaign which Projekttriangle Design Studio developed for fashion house Hugo Boss. Moreover, it is the first store where the new design has been implemented. In Shanghai, the entire concept is dedicated to the subject "Good old Germany..." An over 20 foot high fantasy tree was illustrated by hand and afterwards screen printed onto wooden boards. Detailed carvings and golden motifs decorate the changing rooms. A traditional wooden rubber stamp set consisting of 10 typical German songbirds in life size has been produced. The stamps were directly applied manually onto objects, walls and shelves and function as a label.

Design Agency | Projekttriangle Design Studio **Creative Direction |** Martin Grothmaak **Illustration |** Franz Stämmele **Photography |** Thomas Mardo **Client |** Hugo Boss AG **Location |** Metzingen, Germany

Hugo Boss Orange VIP Store / Showroom

The basic concept the designers developed for the Hugo Boss Orange Monostore in Mannheim, Germany, has been adapted in many stores all over the world. Elaborate graphics and collages of fantastic animals, based on hand-drawn masters, have been implemented – in this case into Hugo Boss Orange's VIP store and showroom in Metzingen, Germany.

Design Agency | TETUSIN DESIGN OFFICE **Design |** Takayuki Senzaki **Location |** Kanagawa, Japan

Yukemuri no Sato

Susukino is now a quiet residential area, but it was the forest of Musashino before. So, the designer drew wall graphics from the basement floor to the second floor to represent the change in scenery from field to forest. He also depicted wood branches on the ceiling to make it feel as if visitors are looking up at the sky in a forest.

Creative Direction | Formavision (Sebastien Agneessens) **Architecture |** Formavision (Sebastien Agneessens) **Graphic Design |** Wall photos shot in Parisian apartment and reproduced at scale 1 with a slight green tint. **Production |** Formavision (Chris Hoover), Site NY **Photography |** Jordan Kleinman **Location |** New York, U.S.

Marithé+François Girbaud Showroom

The New York Marithé + François Girbaud showroom brings contemporary Parisian youth culture to Manhattan through various design and artistic expressions. Designed as a Haussmanian apartment in need of renovation and featuring a mixture of new and salvaged furniture, the space is a destination for intimate concerts by both Parisian and New York bands. The walls of the space are covered with a printed wallpaper from the Parisian apartment used in the film *The Dreamers* by Bernardo Bertolucci, and reproduced at scale 1 in New York with a slight green tint, the iconic color of the brand. The showroom is a direct reference to the early career of fashion designers Marithé + François Girbaud, who made themselves famous by creating show costumes for the French rock and pop stars of the sixties.

Design Agency | TETUSIN DESIGN OFFICE
Creative Direction | Ryuji Takayama **Design |**
Takayuki Senzaki **Photography |** Masahiro Otsuka
Location | Fukuoka, Japan

Hakata GX bld

Graphic design for the display window of a fashion shop.

American Apparel
ê.a.p
Art
garage

Design Agency | TETUSIN DESIGN OFFICE **Creative Direction |** Ryuji Takayama **Design |** Takayuki Senzaki **Photography |** Masahiro Otsuka **Location |** Fukuoka, Japan

Imaginary Water Garden

The "Imaginary Water Garden" is a space for relaxation where people can try water from all over the world. The huge tree not only stands for the circulation of water, but also emphasizes the importance of water in our daily lives.

Design, Photography | Trust in Design **Location |** Vannes, France

B. Institut

B. Institut is an organic beauty parlour. Natural materials and various light sources make the reception area a welcoming place. The large beech display system contains hidden storage and presents the newest beauty products. The computer cut sticker on the shop window produces shadows similar to those that a tree would cast. This enhances the idea of a natural beauty parlour which uses organic products.

privé
U.V

Design Agency | TETUSIN DESIGN OFFICE **Design |** Takayuki Senzaki **Client |** Kimono Hearts **Location |** Kokura, Japan

Camera Studio in Kimono Hearts

Wall graphics for the fitting room where people try Kimono on and take photos.

Design | Jaime Hayon **Photography |** Nienke Klunder **Location |** Barcelona, Kuala Lumpur, Paris and Cologne

Mon Cirque

Mon Cirque is Hayon's vision of a large oneiric metaphor. It is a place that symbolizes hope for artistic freedom, where the designer's work is free to transcend the frontiers of design and art. The classical colorfulness of the circus is replaced by a more serious and elegant contrast between black, white and gold: Cirque de lux. Vases become characters, clowns become lamps, plates are used as canvases, and tables grow legs. A sophisticated ambience replaces the decadent circus. A new world has been born: Mon Cirque. Craftsmanship is a key element to Mon Cirque. A return to the handmade object and the idea of the unique piece is advocated by avoiding the use of any molds to create the forms. Mon Cirque has traveled to Minneapolis, Lisbon, Barcelona, Paris and Cologne. It continues to move on, like a circus should.

Design | Jaime Hayon **Photography** | Nienke Klunder **Location** | Milan, Italy

Hayon at Rossana Orlandi

Three new projects were presented at the Rossana Orlandi Gallery in Milan in April 2008: The Fantasy Collection for Lladro, Hayon's shoes for Camper and his book titled *WORKS* by Gestalten. The creations in The Fantasy collection mirror the perfect fusion between the artistic quality of the Spanish porcelain brand and the playful, fantasy-inspired quality the designer brings to his works. Hayon has also designed a shoe collection for Camper. It is characterized by the fusion of styles, soft and organic forms and the surprising mix of materials that define the personality of the wearer. Jaime Hayon *WORKS*, the first monograph about the renowned creative, presents the full spectrum of his work. Ranging from clear and harmonious to playfully extravagant, all of the examples in this book attest to his inimitable grasp of form, style and color.

JAIME HAYON · PORCELAIN

Design | Ami Suma **Photography |** Marilou Daube **Client |** Yoya **Location |** New York, U.S.

The World of Yoya

Yoya is definitely the most popular baby store in downtown NYC among the city's stylish parents. The owners are aware that the store pushes the boundaries of style even for newborns. So when they approached Ami Suma to decorate the space for a new look, their request was simple and straightforward: "Can you expand the fantasy of Yoya world?" Ami was thrilled because she started painting murals for children hoping to provide a stimulating space to inspire the growth of their sense of whimsy. She felt free when designing this mural.

One thing that Ami felt was important was that her illustration not look like a flat picture; rather, she wanted the whole shop space to exist as a world where things are happening spontaneously. She thought it would be fun for each kid who comes to the shop to find new things each times he visits...

PETIT BATEAU

COBY®
SONY

Design Agency | TETUSIN DESIGN OFFICE **Design |** Takayuki Senzaki
Photography | Kumpei Miyata **Location |** Okinawa, Japan

Kimono Hearts

There is a flower called "Sagari Bana" which is also known as "weeping flower" in Okinawa, Japan.

The flowers, which bloom at nighttime, fall into the river and float to the sea which carries their seeds to other places. The ceiling here is like the surface of water.

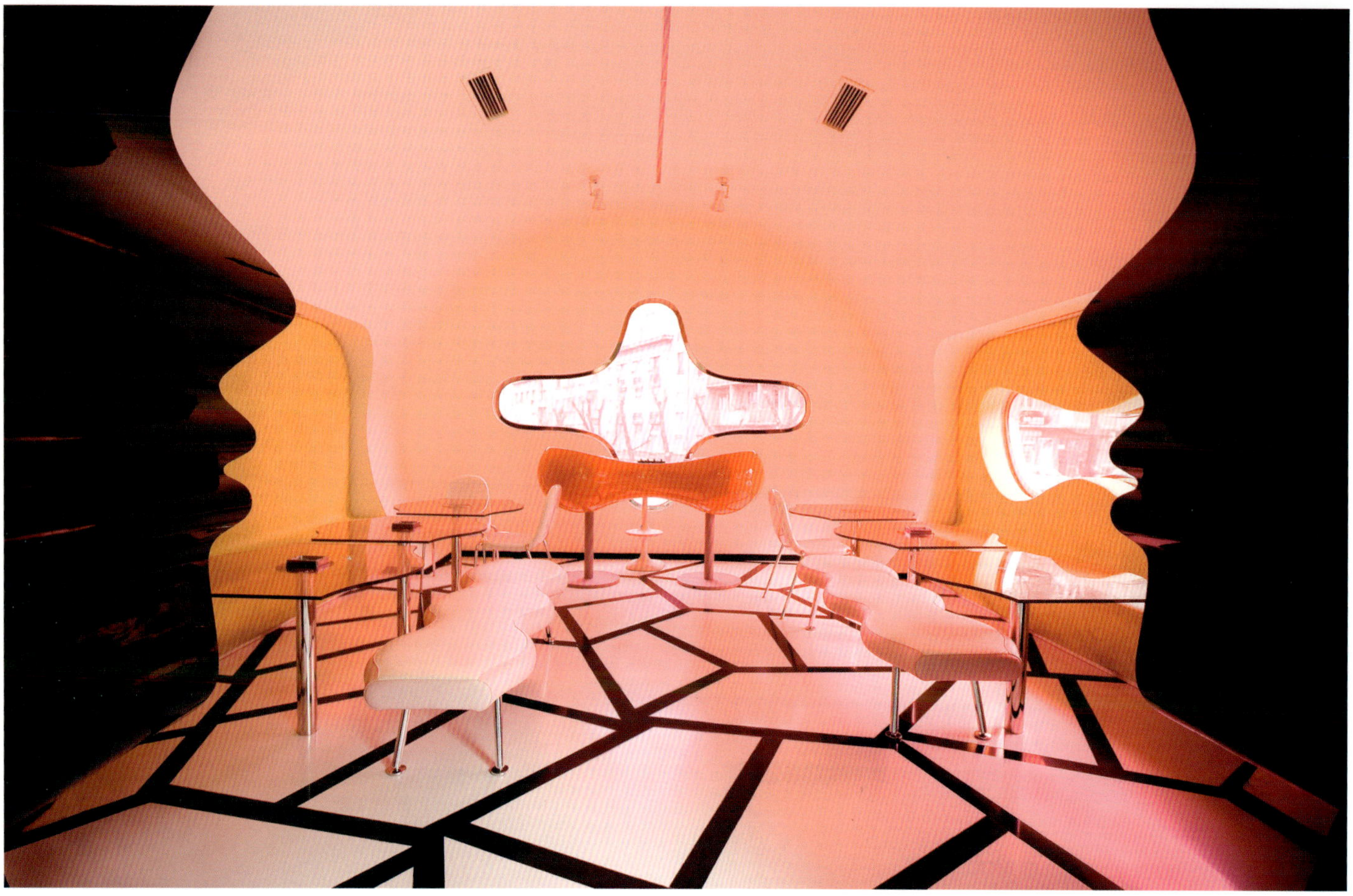

Design | Karim Rashid **Client |** Veselin Jevrosimovic **Location |** Belgrade, Serbia

Majik Café

Majik Café stems from Karim's philosophy of creating a digipop digital nature and a seamless world that excites all the senses. Karim's love of sensuous lines and brilliant hues comes through in this dynamic space. Digital artwork made of RGB LED panels surrounds the bar, making it visible from down the street. The patterned glass bar changes color and mood throughout the customer's stay. The restaurant is brought into the 21st technocentury with a scrolling LED message board which displays customers' text messages. Technorganics blend with humanorganics in the restaurant's details. His and her cutout faces frame the entrance to the DJ Kreemy booth lounge area. Profile mirrors reflect patrons in the restrooms with data resonating from the eyes. The exterior is finished in polished stainless steel and mosaic tile. Icon shaped windows and decorative ancient Serbian patterns are updated to create a kaleidoscopic, engaging and inspiring radiant space for lounging, dining and drinking. The restaurant's message is east meets west digital freedom, where technology and new social behaviors are the global proponents of a new singular seamless world.

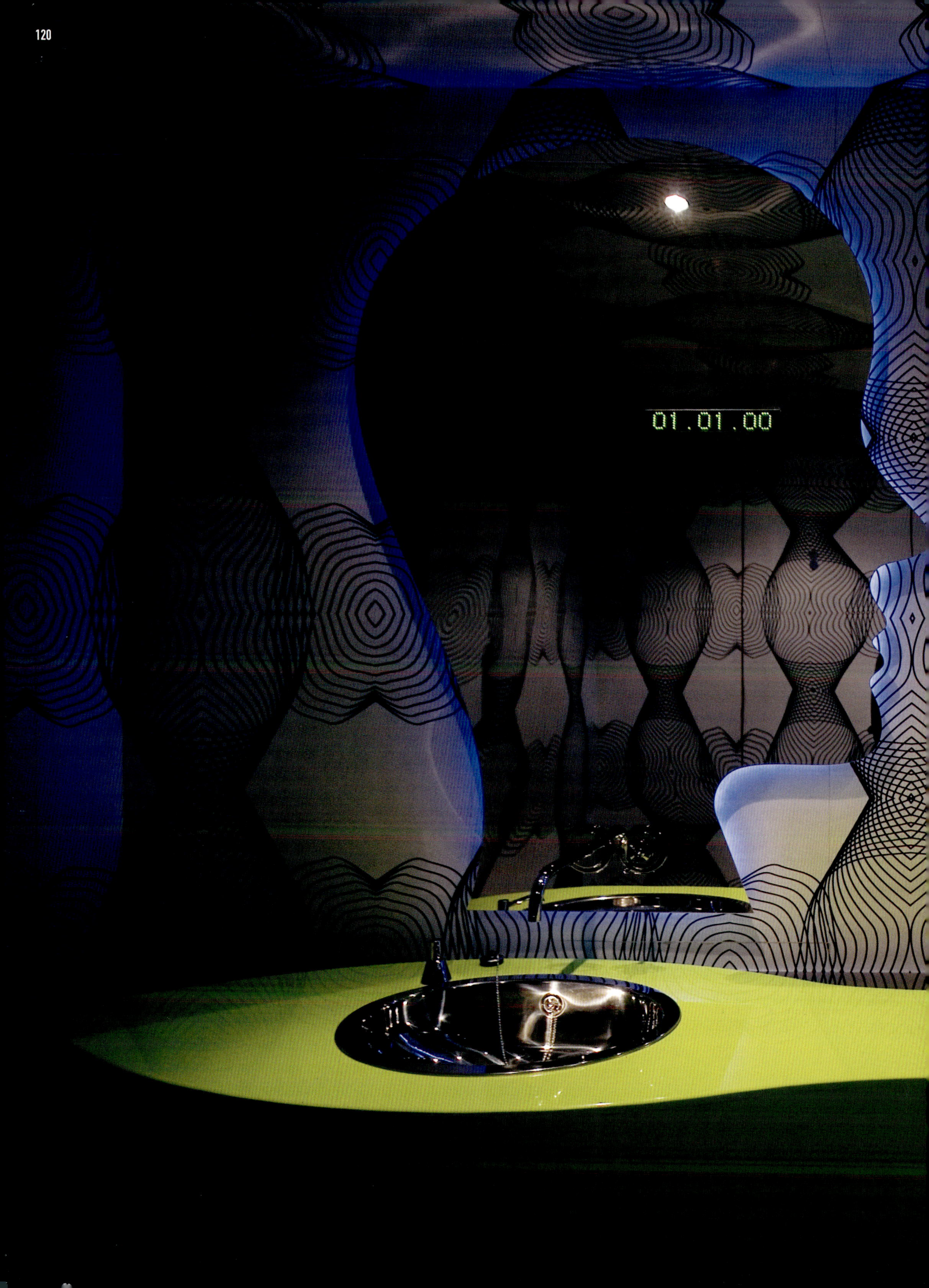
01.01.00

23.12.07

Design | Mr. Important Design **Photography |** Farrah Karapetian **Location |** San Jose, U.S.

Motif

The mix of vintage and new, the desire to push the use of materials, the need to transform the interior from restaurant to club ambience and the evocation of a kind of Asian feeling of dark glamour were the objectives in the design of Motif.

Mr. Important Design mixed things up by using futuristic materials such as stretched glossy polymer film to provide an undulating ceiling of reflected color and patterns in the DJ area. They used cutting edge Asian furnishings with traditional materials from Kenneth Cobonpue coupled with the modernism of Konstantin Grcic intermingled with the '60s and '70s glam of Tommi Parzinger style pendants and sleek black linear silk shades.

The designers pushed the use of materials. Working with Eva Menz, they suspended seven thousand pieces of black glass forming two huge leaf shapes above the downstairs restaurant and lounge (Eva Menz). Hundreds of feet of black and silver chain were hand worked into a bold floral tapestry (art provided courtesy of Amy Butler) that envelopes the space without suffocating it. Yards and yards of thick black silkrope soften the mood of and provide a pattern to the institutional cinder block construction of the building's interior.

Mr. Important Design used lighting to help transform the interior. Extensive and saturated use of color changing LED lighting transforms the intentionally monochromatic palette of Motif's interior into a heady polychromatic club experience later in the evening. One of the main challenges was to make the switch from restaurant to club with just the ambient lighting. The designers created an entirely monochrome interior that would be appropriate for both daytime and evening functions. In the daytime, the restaurant's monochrome furnishings are bathed in a warm amber glow from Edison bulbs within black silk shades. Later in the evening, the full effect of the extensive l.e.d. system floods the interior with polychromatic lighting and slowly changing effects herald the beginning of serious nightlife.

Creative Direction | Sebastien Agneessens **Architecture |** Formavision, Sebastien Agneessens **Graphic Design |** Rostarr **Production |** Formavision (Chris Hoover, Shinya Nakamura), Genart, Wieden Kennedy **Photography |** Jordan Kleinman **Location |** New York, U.S.

Starbucks Salon

The Starbucks Salon is a nomadic temporary coffeehouse concept combined with a gallery and performance venue, which features daily performances by musicians and artists such as Jose Gonzalez, Aimee Mann, Run DMC, Jim Carroll, House of Diehl and Lansing Dreiden. While the first floor was conceived as an artistic coffee shop and intimate performance space, the second floor exhibits a slanted gallery installation that aims to represent the new perspective on contemporary culture brought by increasing corporate patronage.

Marshall

Design | Trust in Design **Photography |** Stéphane Chalmeau
Location | Nantes, France

La Suite 21 Club

The red lacquered and perforated ribbon wraps up all the club's functions (bar, VIP lounge, DJ booth, etc.). Furniture is set up within a large lounge and indoor terrace. The sofas enable various comfortable positions allowing the patron to view the show and chat with friends. The graphic ribbon is like a large fresco with different levels of meaning, from the lace pattern to small provocative stories and hidden jokes.

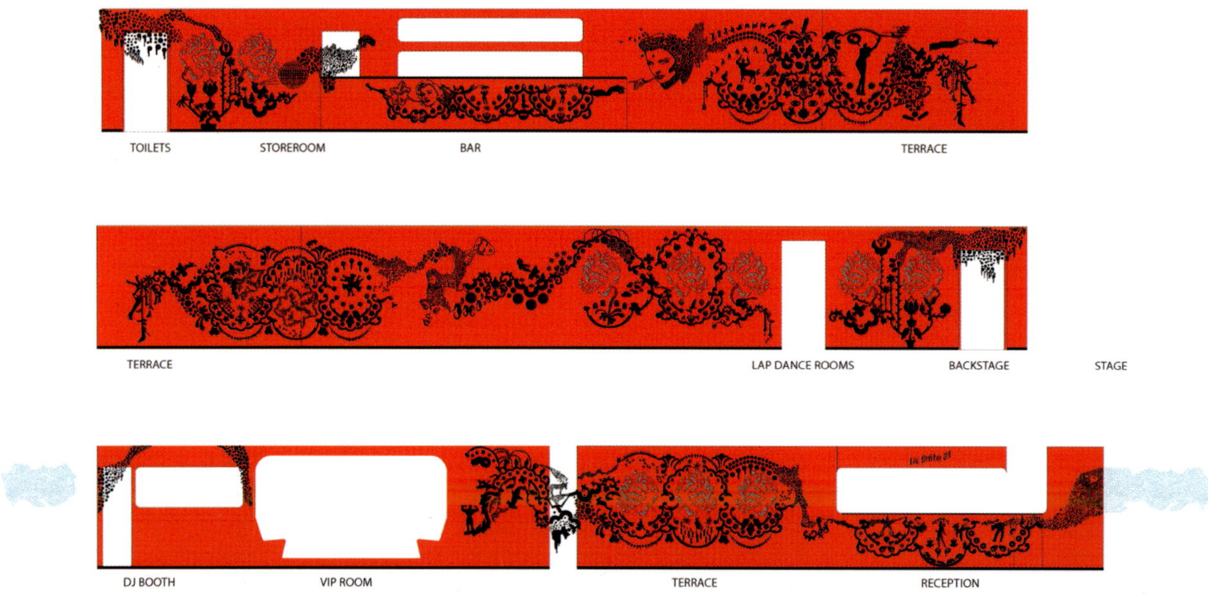

Creative Direction | Stuart Taylor **Design |** Spread Design
Illustration | Ben Allen **Photography |** Alistair Lever
Client | Urban Leisure Group **Location |** London, UK

Graphic Bar

The design of the Graphic Bar was derived from a brief based around urban art and the London urban landscape. The Graphic Bar is situated overlooking Golden Square in Soho and attracts a trendy mix of people within creative industries. It transforms from a relaxed morning/day venue for breakfast and brunch into a cocktail bar featuring an array of DJs.

The design is tailored to integrate a variety of urban artists who change the space every six months by placing their artwork onto the walls (starting fresh with a blank canvas), allowing the bar to grow and evolve with trends within London. The seating and lighting will be altered, if required, to allow the art to remain prominent, enabling Graphic Bar to remain at the forefront of the urban scene.

LOVE
graphic

Design, Illustration | Alessandro Suzzi **Photography |** Giovanni Albore
Location | Bisceglie, Italy

Moulin Rouge

Moulin Rouge is a bar developed by a lover of classical Parisian clubs.

The idea was to recreate the same atmosphere, but with a more modern and current style.

Bright colors and attractive women make this area suitable for a young customer who wants to enjoy a drink in a small moulin rouge that evokes the spirit of party.

Design, Illustration | Alessandro Suzzi **Photography |** Giovanni Albore
Location | Carpino, Italy

Gran Caffè del Corso

Gran Caffè del Corso is a classic café.

The client required a simple and elegant design in order to create a minimal interior with floral elements.

Design, Illustration | Alessandro Suzzi **Photography |** Giovanni Albore
Location | Bisceglie, Italy

Bakery Café

Bakery Café is a project that focuses on color choices in relation to form.

The client wanted a dynamic environment with soothing colors in order to create a lasting pleasant environment for the customer.

Alessandro Suzzi took care of the entire design and implementation of the corporate image. The result is a sober and elegant club.

café
Bakery

café
Bakery

Design, Illustration | Alessandro Suzzi **Photography |** Giovanni Albore
Location | Trani, Italy

Spazio O.F.F.

Spazio O.F.F. is a lively club located in a large industrial building. The large bar needed decor that would inspire customers to relax. The dynamism of the wall imagery breaks up the square shapes that characterize the club.

Creative Direction | Charlotte Henson **Illustration |** Pomme Chan **Client |** Topshop **Location |** London, UK

Topshop

Topshop came up with the idea of having eight illustrators create graphics for their t-shirts as well as exhibit work in the store. Wall paintings in the stores (both in Oxford Circus and Manchester) showcase unique works and communicate the store image to customers.

The illustrative concept comes from the brief allowing the illustrators to draw anything that corresponded to key phrases such as "sweet but sinister," "bat for lashes," "ominous yet beautiful," etc. Pomme's illustration is composed of graceful roses and stems in a darker style than she typically creates. The image represents women who are sweet like roses yet are not innocent.

ILLUSTREIGHT

Design, Illustration, Photography | Camille Rousseau **Collaboration |** Pia Fairhurst, Zizzi's Head of Design **Location |** London, UK

Zizzi Charlotte St

The flowing scene is inspired by both the local London area and the far away lands of rural Italy – the familiar sights of Charlotte Street and Percy Passage frame the gastronomic delights of the Italian countryside which spill out of the setting. The two women exchange glances across this vista of London and Italy – the question remains, who is awake? And who is dreaming of whom?

Elements are composed as if they were linked to each other. The streets of London are framing the dream.

CHARLOTTE
STREET

Street Specials
up with a touch of marscapone
ugh sticks
ana
lla, rich tomato sauce and garlic
h grana padano cheese
ed with herby tuscan potatoes and rocket
n amazing melting middle served with
artori
Small glass 4.50
BadCat
Marshall
Peavey
SAVE
FISH
EAT
CHIPS
CHA
STE

Design, Illustration, Photography | Camille Rousseau
Collaboration | Pia Fairhurst, Zizzi's Head of Design
Location | Winchester, UK

Zizzi Winchester

The illustrations are derived from the space itself, which was once a printing company. The work is an exploration of typographic iconography from its genesis, to the advent of the printing press and beyond. The characters depicted all played crucial roles in the development of printed and hand rendered typography. The illustrations are all derived directly from type. Capital letters in ancient manuscripts were traditionally called viticuloe or vignettes. More recently the term has been used to describe all kinds of wood cuts or copper plate engravings. The word is French and is homonymic with the Latin word *viticula*, meaning a little vine or vine shoot.

1772
Vol. IV.

HAMPSHIRE
1772
ANDERSON
EXTRAORDINARY
Hampshire Chronicle

Winchester Cathedral Thanksgiving
HAMPSHIPE CHRONICLE.
MONDAY, JUNE 9, 1823
Expreſs from London.

NE 9, 1823
Expreſs from London.

Vol. IV.

MONDAY, July 8, 1776.
[Numb. 203]
The past week has been productive of no occurrence, either foreign or domestic, of sufficient public interest to claim particular notice.
772
Vol. IV.
HAMPSHIRE

Creative Direction | Adam Gardiner **Design, Illustration |** Janine Wurfel, Radek Wojcik **Photography |** Qube Konstrukt & Shannon Mcgrath **Client |** Pancake Parlour **Location |** Melbourne, Australia

Pancake Parlour

Creation of venue imagery for the newest addition to a family-owned chain of restaurants. With a rich and eclectic history of interior graphics for previous venues, the brief was to continue this tradition while bringing a fresh approach and visual style to the graphics.

The scope of the brief included a large area of wall space to be covered, including an area requiring a 3-dimensional build. The designers' response was to create a series of flowing and interconnected illustrations for the main room, a custom designed wallpaper for the function room and 3-dimensional illustrated panels for the entryway.

Design | Studio Makkink & Bey **Photography |** Stijn Brakkee
Location | Milan, Italy

Stand Milan 2009

In Fuori Salone 2009, PROOFF displayed its latest work and innovative designs for the working space of tomorrow. Among the exhibits, there are Earchairs and a Work Sofa designed by Studio Makkink & Bey.

PROOFF
PROOFF
PROOFF
PROOFF

PROOFF
PROOFF
PROOFF
SV

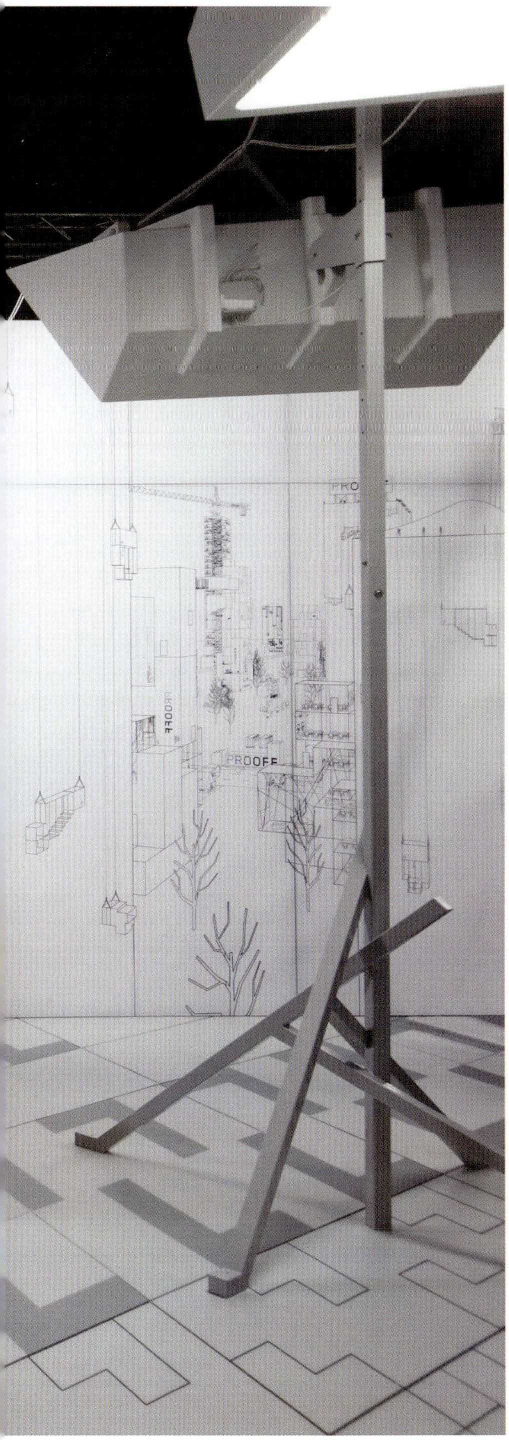
PROOFF

Design | Dizel&Sate **Photography |** Fredrik Sweger **Location |** Stockholm, Sweden

Siemens

Graphic interior design for Siemens in Stockholm, Sweden.

Design | Dizel&Sate **Photography |** Fredrik Sweger
Location | Stockholm, Sweden

Suite 809

Graphic interior design for one off Suite at the Hotel Anglais.

The view from the room made the designers start thinking of throwing paper planes towards the park. That's how they came up with the idea for this artwork. It simply symbolizes the process of folding paper planes. The most basic paper plane would only take at most six steps to complete. Leonardo da Vinci is often cited as the inventor of paper planes, although this is debatable since the Chinese invented both modern paper and the kite. The earliest known date for the creation of modern paper planes is said to have been in 1909.

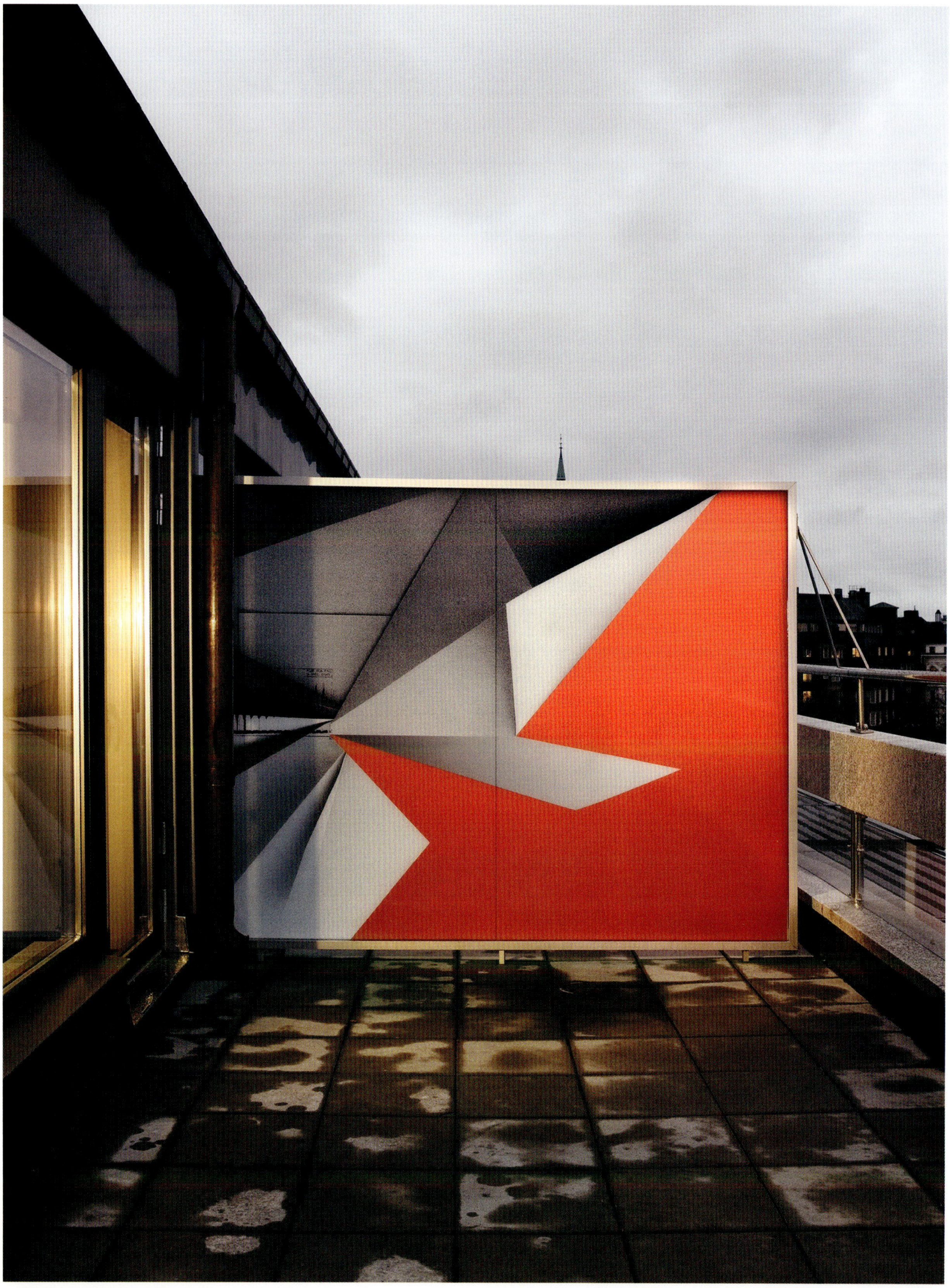

Creative Direction | Bitten Stetter, Jutta Südbeck **Illustration |** Bitten Stetter, Jutta Südbeck **Photography |** Ann-Kathrin Kampmeyer **Location |** Fitting Forward concept store, Hamburg, Germany

Tingeltangel – A Honky-Tonk-Stroll down the Whoopee

Ladies and Gentlemen, welcome everyone! Come in and enjoy the spectacular!

Without further ado, please welcome the zeitgeist-circus of fashionable and colourful clownish freaks and brand new attractions! We present design-tamers, knife-throwing silhouette-tailors and impulsive magicians.

Step right up, sit right down in our circus tent full of funny blockbusters from bumper cars to Megadrome, from trapeze artists to fire eating snake-charmer. This Tingeltangel juggles ferris wheels and circus horses, dances on the high wire with the ballerina and hits the high striker right through the roof. Visitors will be thrilled by design attractions, style-acrobats, and by the bizarre and the odd.

Sensations, Attractions, Mutations. Experience our fantastic atmosphere with unbelievable creation-gladiators, multi-talents in any imaginable aesthetic-acrobatic facet.

Triumphant on stage or being upstaged – who will win the stylish jackpot?

FITTING FORWARD

FITTING TALES

Creative Direction | Bitten Stetter, Jutta Südbeck **Illustration |** Heliana Rotariu, Romania **Photography |** Gulliver Theis **Location |** Fitting Forward concept store, Hamburg, Germany

TalenTiere - Zoo of Uniqueness

The "TalenTiere" are gathered under the unique-species-program and live together with rainravens and elephantjumpers, dinosaurnecklaces, parrotjackets, tiffanycats and mulesideboards in the zoo of uniqueness. This zoo opens its gates for wondrous textile-animals and snappy design.

On a daily basis, unseen and uncared for "TalenTiere" disappear from our earth due to seasonal rapidity, neglected design-quality and elimination of design-habitats. The zoo of uniqueness is trying to halt the upheaval. With special absurdity-protection-programs and wild space- and staging-projects it is speaking up for the rescue of the threatened "talented animals."

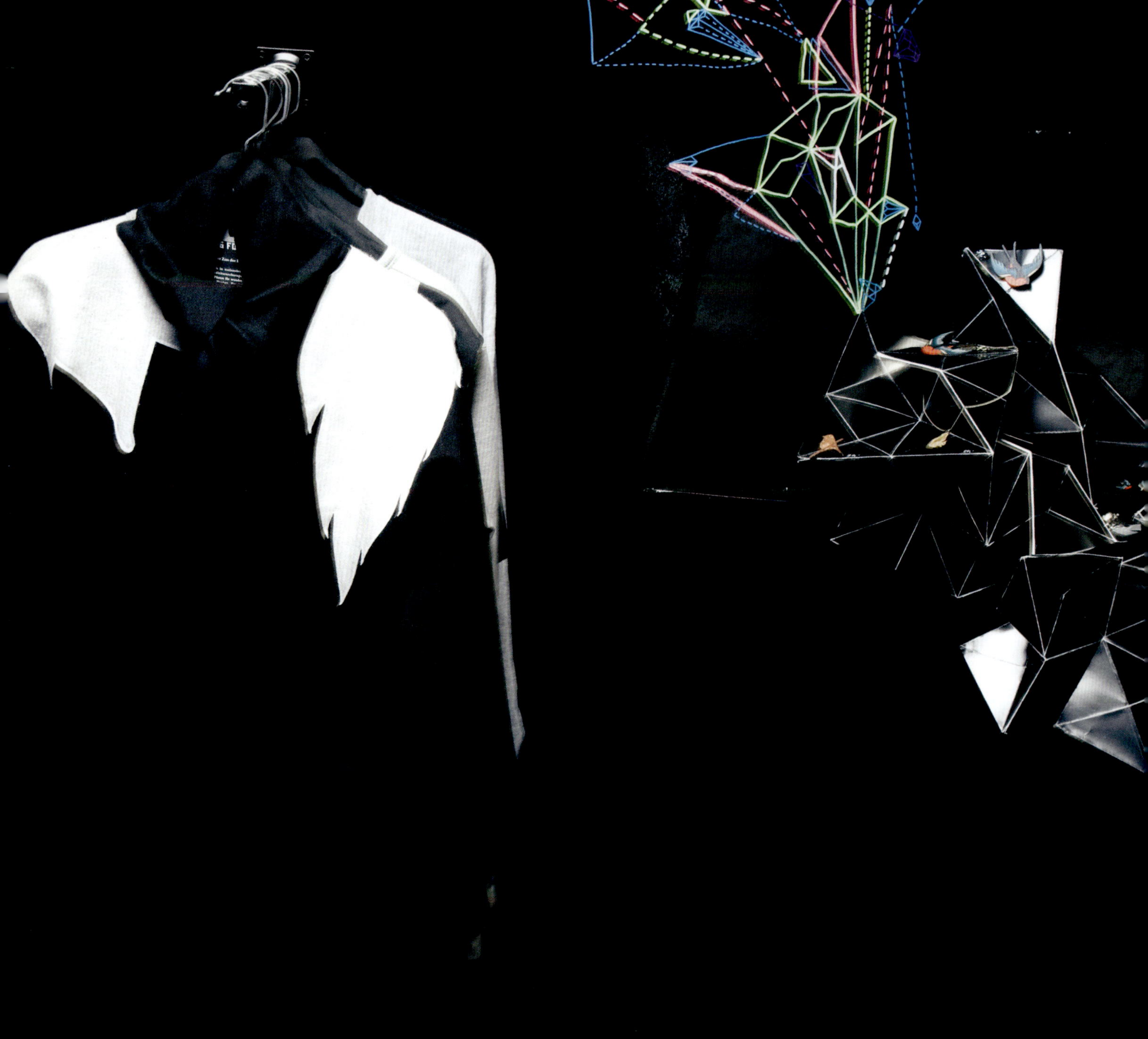

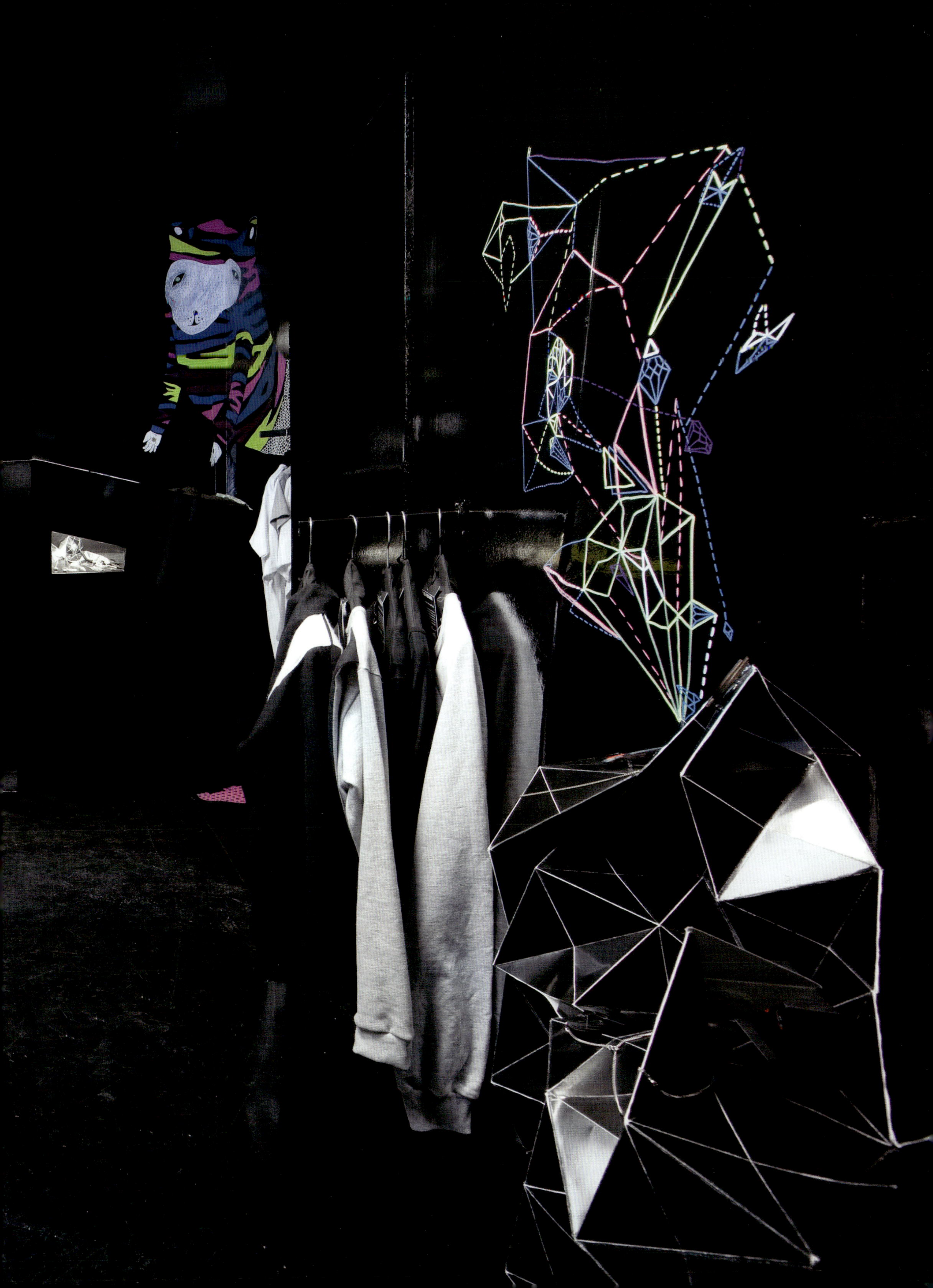

Creative Direction | Bitten Stetter, Jutta Südbeck **Illustration |** Paul Steen, Sweden **Photography |** Gulliver Theis **Location |** Fitting Forward concept store, Hamburg, Germany

Darwin's World - in Search of "Hans im Glück"

We are Hans on his tramp of luck in the Brothers Grimm tale. Illustrations within the space correspond to Darwin's theory of the adaptation of the biotope via variation and natural selection. The road depicted on Han's journey is surrounded by a diversity of life.

"Hans," said the Master within the tale. "You have served me faithfully, and as was the service shall be the wage." The Master then gave Hans a chunk of gold the size of his head. However, throughout the course of the tale, Hans continues to swap the gold for objects which seem more useful to him, until he finally reaches his destination. What good is gold if you can exchange it for a horse to ride on instead of walking? What good is a horse if you can have a cow? Do achievement and talent matter? Is the journey in fact the reward? Within the store, we visually explore heroism and the surrounding world, search for luck, consider parallel currencies, evaluate and resist embedded values, and swap prizes to find the best objects to aid in solutions within our lives.

Creative Direction | Bitten Stetter, Jutta Südbeck **Illustration |** Eklatant///Enrico Nagel, Germany
Photography | Ann-Kathrin Kampmeyer **Location |** Fitting Forward concept store, Hamburg, Germany

On Your Marks, Get Set, Role Model – The MegaHeroSuperAlarm

Here we go, go – go, as we step out of the normal sphere and go crazy about superlatives, superheroes and superstars to celebrate ourselves in the MegaHeroSuperAlarm.

We are brave and strong-minded, gladiatorial, we interfere without fear and blame, are present for each other, even without reward. We impress inevitably by superiority and achieve exceptionality for us and sometimes for others as well. We actively override financial stagnation and boost consumerism without fear. We are knights in shining armour, workaday-heroes. Gladly embracing a position in front of the camera, we do not dread catastrophes nor storms of flash-lights. We like to be in the first row.

We get underestimated, sometimes we stay unrecognized – no wonder, as we disguise ourselves perfectly or even turn invisible if a situation becomes dodgy. We are always fast and ever present – our resources are not limited.

We are super stress resistant, no Career-flash discourages us. We are superhero and fan at the same time; we are enthusiastic devotees of ourselves, our powers and our achievements and even like to talk about it. We costume ourselves with frock and pseudonym and mostly keep our true identity a secret. We are digital, we are human-machine 02. We gather anonymously in fan clubs, to report and count our own heroic acts as well as our followers. We are fans of having and being fans.

Just be brave. To be a hero suits you well.

MAX GRÜTER
HOUSTON
I AM A PROBLEM!

WHO'S
NEXT?

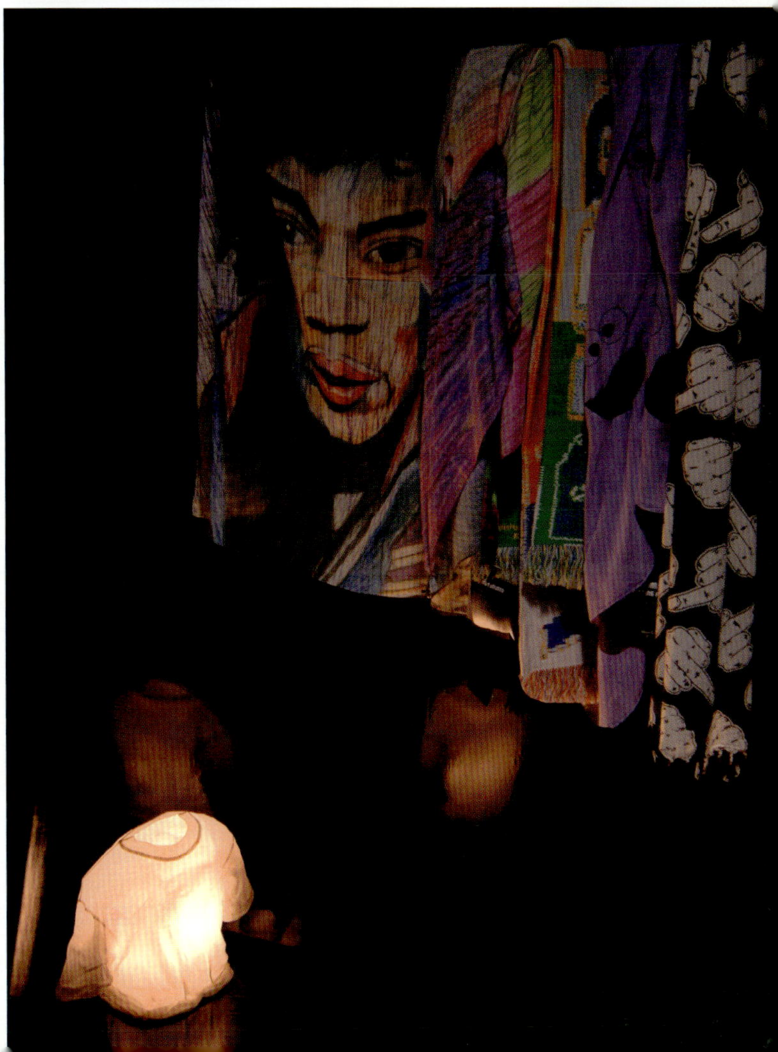

Creative Direction | Bitten Stetter, Jutta Südbeck **Illustration |** Katja Musenberg, Germany & black "terrorist-balloons" by Vlad Nanca, Romania **Photography |** Ann-Kathrin Kampmeyer **Location |** Fitting Forward concept store, Hamburg, Germany

Urbansinn – The Neurotic Metropolis-Massif

Anonymity, Abnormality and Normality, Identity-lust and-loss run through the current city-map for Fitting Forward. A unique and haunted urban odyssey navigates through the metropolis traffic jam. This design-scenario dashes through concrete walls, hides in remote tunnels, crosses urban style-junctions, ignores colour-lights, brakes for the avant-garde and knocks over all design stereotypes.

Fitting Forward is a laboratory of modernity, a place of compression and acceleration, and a place of disguises, urban camouflage and luminous colour battles.

This is the call for a fashionable rampage.

Creative Direction | Bitten Stetter, Jutta Südbeck **Illustration |** Tasek, Germany **Photography |** Gulliver Theis **Location |** Fitting Forward concept store, Hamburg, Germany

Parlour-game grinch - eenie meenie meinie moe!

Overtrumping, gambling, bluffing, faking, kicking-out team-mates, sitting-out one round, beginning again, winning or losing, and passing the buck to someone. It's a "dog eat dog" world that annoys us!

Within this world, we create a new theme and break all the rules of the game. We open the doors for all the party-poopers who dare to do or say something new and meaningful—those who attract others' attention, want to make a difference, or long for new rules, no longer wishing simply to join a flawed game.

So, join us in the launch of Fitting Forward painting fresh tones on shirts and forming avant-garde alliterations out of fashionable metaphors, and for TASEK's wordplay.

Altogether different, we are open, unprejudiced, and dedicated to finding and presenting that which is valuable but overlooked, rendering these gems tactile in phonologically playful clothing, game-revitalizing accessories, and radically groundbreaking stories.

ENRAUSCHEN

[Auf alten Pfannen lernt man kochen!]

FITTING TALES
FITTING TALES
FITTING TALES
FITTING TALES
FITTING TALES
FITTING TALES
FITTING TALES

Creative Direction | Sebastien Agneessens **Architecture |** Jeroen de Scrhijver, Ellen Depoorter **Graphic Design |** Mai Kato, Jonas Hjertberg **Production |** Chris Hoover **Photography |** Jordan Kleinman **Location |** New York, U.S.

Reebok Flash

Formavision, which conceived and developed the concept store "Reebok Flash," took inspiration from Vorticism, an English arts movement from the early 20th century noted for its dynamic interpretation of Cubist and Futurist principles. Combining Vorticism's vibrant aesthetic with an assortment of cultural cues ranging from *Purple Rain* to *Miami Vice, Flash Dance* to *Thriller*, Formavision sought to capture the pop spirit of the '80s in order to create the ideal environment in which to reintroduce these classic styles from Reebok. The design of the space plays with depth and perspective, tricking the eye by extending three dimensional shapes into distorted graphic patterns, a camouflaging technique reminiscent of the Royal Navy's dazzle ship graphics from the First World War. Their intent is to provoke and perhaps confound visitors by making them feel as if they are stepping into a poster rather than a store.

THE pump
TM
SQUEEZING YOUR TONGUE
CAN PROTECT YOUR ANKLES
Reebok
Reebok

Design | Dizel&Sate **Photography** | Fredrik Sweger **Location** | Stockholm, Sweden

Vasakronan

Vasakronan is the largest real estate company in Sweden. In collaboration with Murman Architects, Dizel&Sate created the graphics for their office.

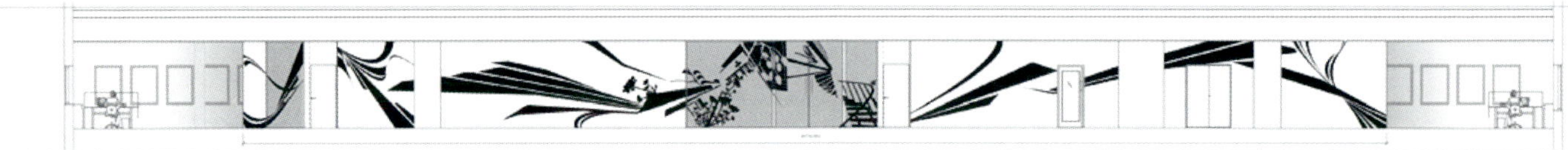

ELEVATION B-B MITTKÄRNA HUS C

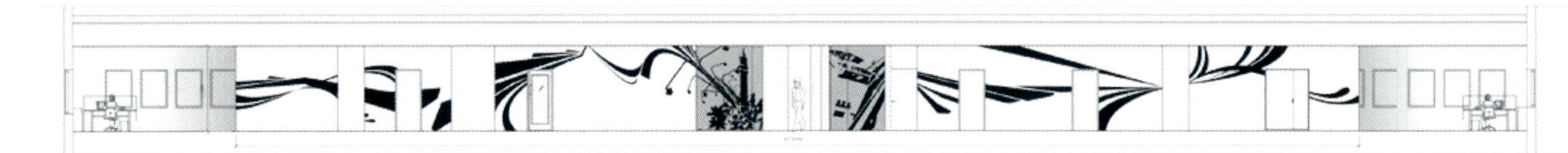

ELEVATION A-A MITTKÄRNA HUS C

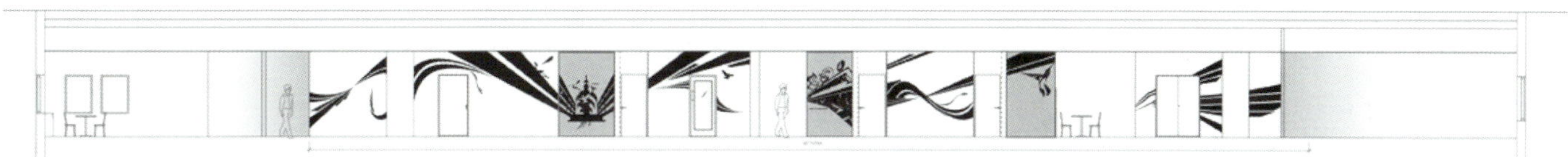

ELEVATION B-B MITTKÄRNA HUS B

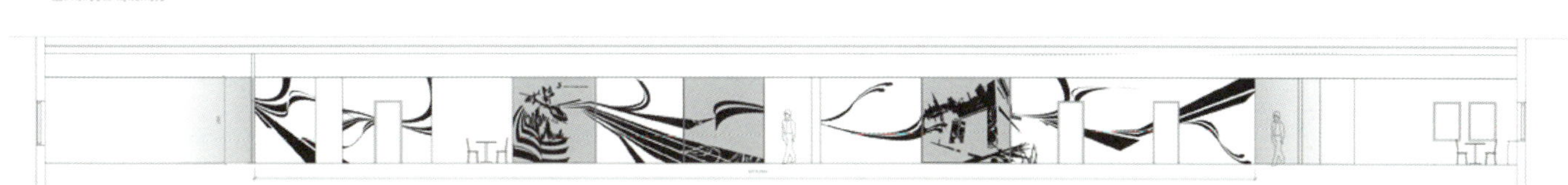

ELEVATION A-A MITTKÄRNA HUS B

Illustration | Carlos Ortega **Photography |** Juan Marcos Castañeda **Client |** Coca Cola FEMSA **Location |** Mexico City, Mexico

Espacio C Mixcoac

The first training facility called Espacio C ("C" for the initials of Coca Cola—Training, Quality, Commitment and Creativity in Spanish) is located at the Mixcoac Distribution Center in Mexico City. It includes three separate rooms that can be joined together to use for common activities, as a media center, and to hold a snack bar, souvenir store, data and electrical connections, and a storage space. The brief also required the incorporation of elements from the different brands of Coca Cola FEMSA and the reinforcement of company values within the space.

These spaces are surrounded by a wall that acts as a sound and visual barrier that isolates the training facility from the city and the rest of the areas of the plant. The outer wall was conceived as a continuous surface on its outer side that folds into the floor and ceiling. The walls and ceiling have a Coca Cola Red epoxy paint coating and the floor is finished with a high resistance acrylic finish. The graphics on the exterior create a composition based on the Coca Cola logo formed by a continuous wave of bubbles, animals and plants. The interior walls are covered with a special anti-graffiti coating that turns the full surface of the wall into a dry erase board and at the same time encapsulates the vinyl graphics in order to protect them from damage.

open!
integridad, austeridad

somos

Design | Via Grafik **Location |** Amsterdam, The Netherlands

Tommy'z Toko

Tommy invited Via Grafik to Amsterdam to repaint the walls of his barbershop. For the occasion of the event "Kersvers," they treated the walls like single canvases that are linked together by the overall theme of hair.

Design | Pandarosa **Photography, Client |** UDK (Universitat der Kunste)
Location | Berlin, Germany

Design Transfer

Installation piece that aimed to distort the perspectives of above & below by altering the foyer into a "reflective" space where perception is skewed & inverted.

This approach aims to visually represent Pandarosa's migration from the southern hemisphere (Melbourne, Australia) into the Northern hemisphere (Berlin, Germany), and examines both contrasting & similar topographic features found in both locations.

Design | Dizel&Sate **Photography |** Fredrik Sweger **Location |** Stockholm, Sweden

Optilon

Optilon is a supply chain management agency in Stockholm, Sweden. The concept is inspired by the fashion industry and specifically, by the style of Japanese designer Yohji Yamamoto. Futuristic cuttings play the masculine against the feminine. Yamamoto's asymmetric and poetic draped dresses informed the use of a black and white base with an unusual punctuation of bright colors. The chosen palette and form also resemble the landscape of an archipelago.

SCM/APS
45
29
36
61
Optilon

Design | Dizel&Sate **Photography** | Fredrik Sweger **Location** | Stockholm, Sweden

Svartensgatan 7

Dizel&Sate has developed the interior design concept and graphic profile for Svartensgatan 7, a salon named after its street address on the historic Mosebacke square in Stockholm. The idea was to capture transparency and light in a rather dark space with limited natural light. The design is futuristic with a color palette of subtle, natural colors juxtaposed with harsh fluorescent lighting. Glass surfaces, light panels and walls are covered by illustrations and typography that capture fictional environments and people in motion.

The custom components, such as shelving and the reception desk, are made out of lacquered naked MDF. Oblique angles, inspired by the number seven, appear repeatedly throughout the space; for instance, in the sloping sides of the reception desk. The lighting is clustered in the center of the ceiling, with tangled cords creating a sense of randomness. Svartensgatan 7 also features vintage furniture by Finnish architect Eero Saarinen.

The new interior includes a retail area, a reception area, seventeen work stations, two shampoo stations and a lounge, which together take up a total area of 260 sq meters.

Design, Illustration | Alessandro Suzzi **Photography |** Giovanni Albore
Location | Trani, Italy

Lunch Bar

Lunch bar is an explosion of geometries.

The project started with the intention of creating depth cracking the wall.

The space is enlarged by an optical illusion in a modern and elegant style.

Design | Esther Stocker **Photography |** Andy Keate **Courtesy |** Museum 52, London
Location | London, UK

What I Don't Know about Space

Masking tape, foam core and emulsion paint on wall, 3.69m x 10m x 2.8m, 2008.

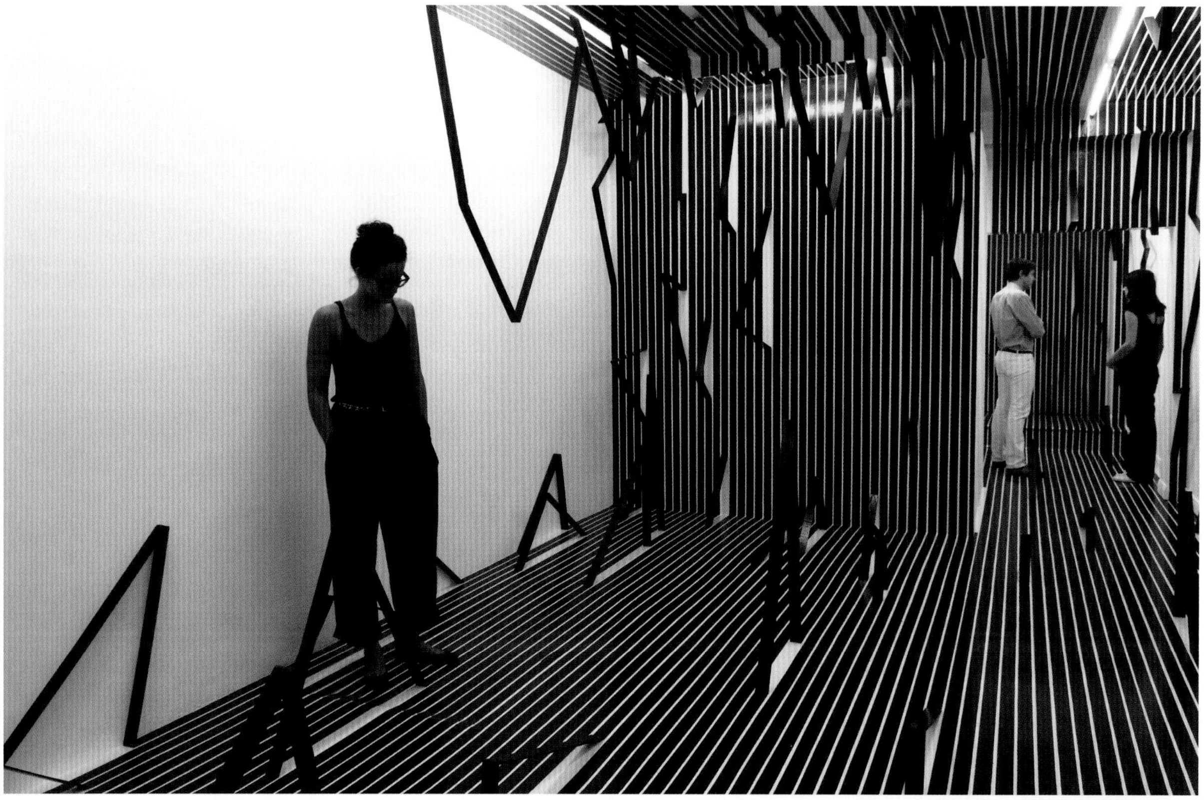

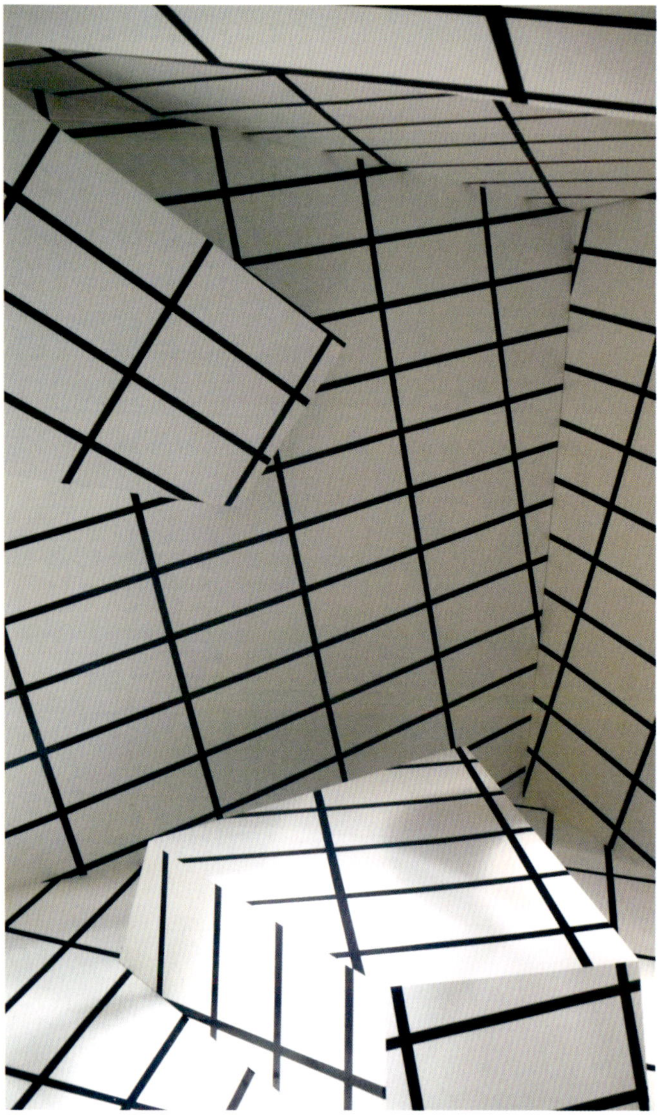

Design, Photography | Esther Stocker **Courtesy** | Lift Off 2008, Eindhoven **Location** | Eindhoven, The Netherlands

The Truth of Basics, Resetting the History of Living between Four Walls

Masking tape and emulsion paint on wall and wooden elements, 241cm x 359cm x 345cm, 2008.

Design | Esther Stocker **Photography |** Sacha Georg, Paul Julien Robert **Courtesy |** CCNOA, Brussels **Location |** Brussels, Belgium

Abstract Thought is a Warm Puppy

Wood, emulsion paint, 10.75m x 12.10m x 3m, 2008.

Design | Esther Stocker **Photography |** Andy Stagg, Esther Stocker
Courtesy | South London Gallery, London **Location |** London, UK

Beyond These Walls

Foam core, tape and pins on wall, 2.92m x 6.85m x 5m, 2009.

Creative Direction | Magnus Zwyssig **Design |** ZMIK **Photography |** Eik Frenzel **Client |** iart interactive ag **Location |** Basel, Switzerland

iart

iart interactive moved their offices to a new place inside a building from the seventies. This spacial intervention in the main circulation area visually enlarges the narrow corridor by using the medium of anamorphosis: looking from 5 fixed vantage points, the wireframe-drawings show real and imaginary rooms behind the walls. Besides these points of view, the graphic merges into abstract lines. For iart, the intervention is also an allegory for the quest for new perspectives.

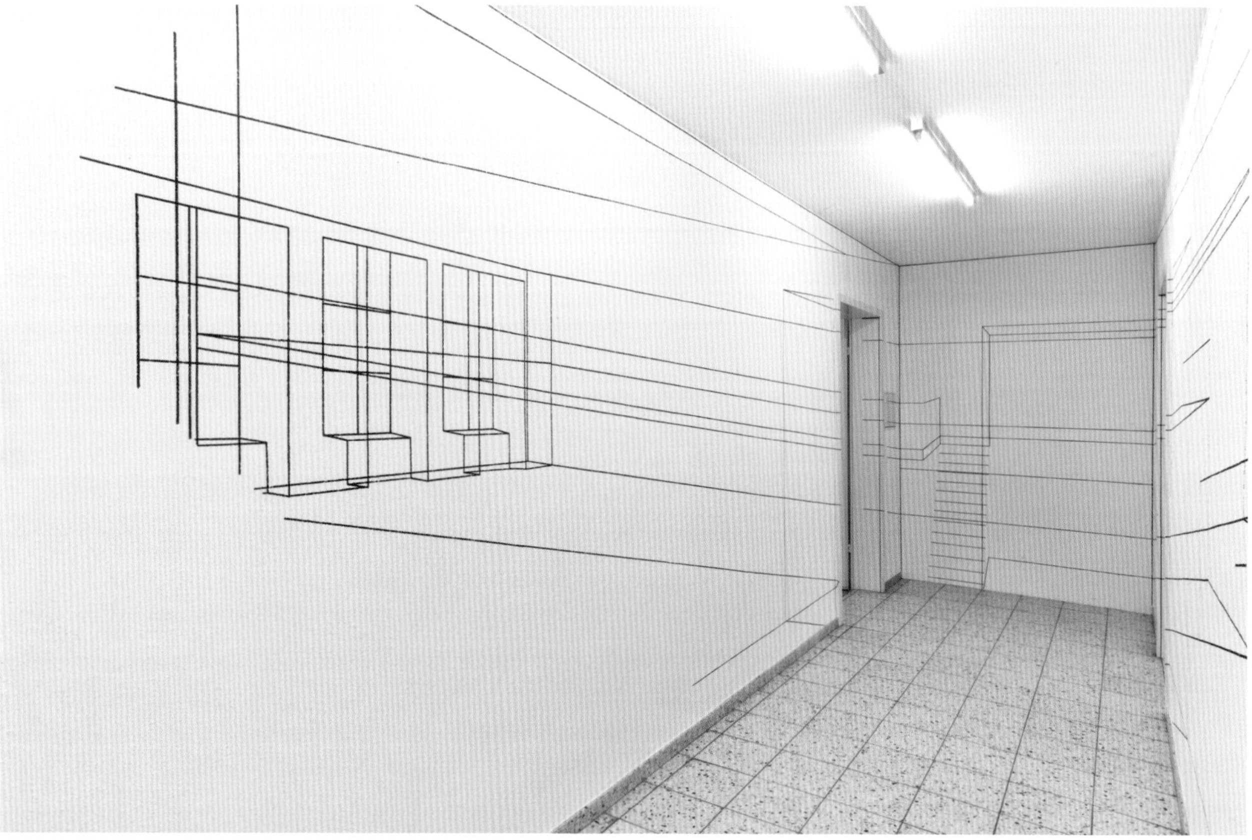

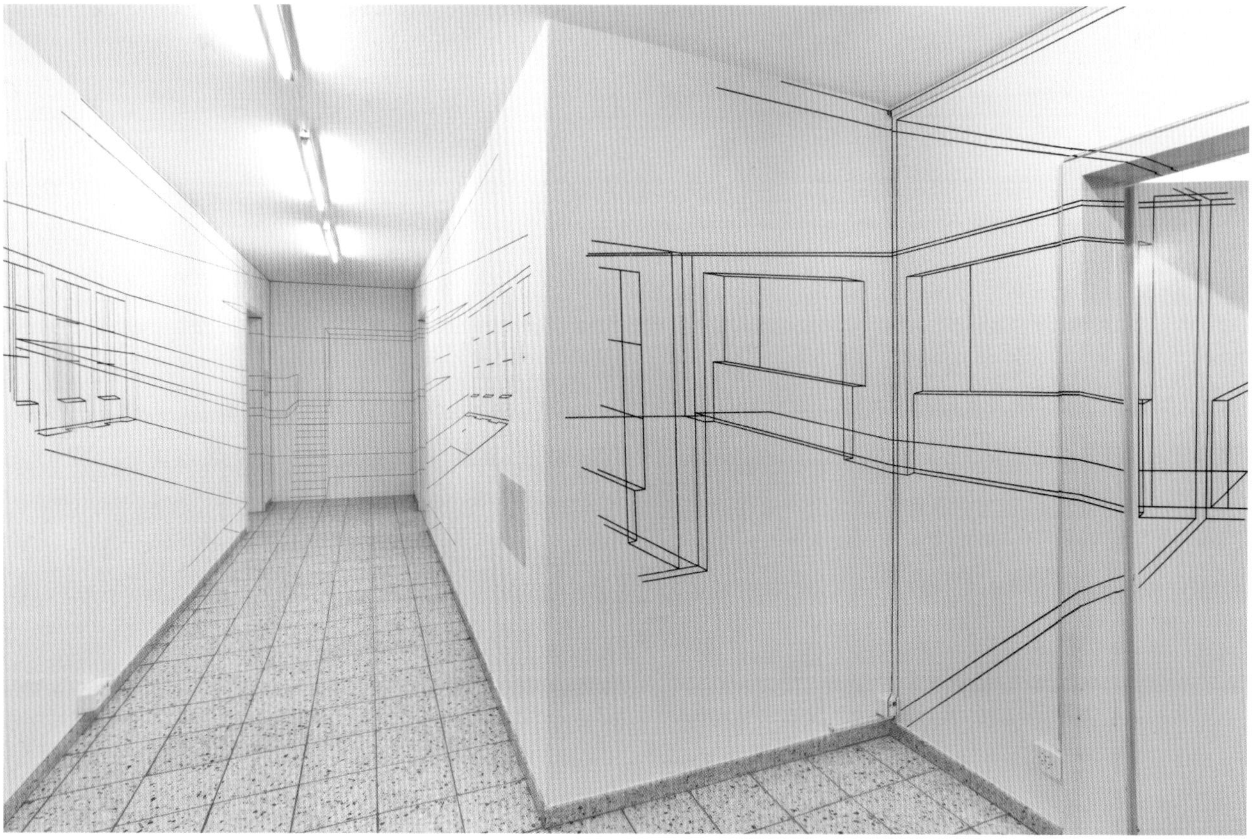

Design | Rebecca Ward **Location |** New York, U.S. & Toyko, Japan

Kate Spade New York

Installation design for Kate Spade New York, Broome Street Flagship (New York) and Aoyama Flagship (Toyko).

Design | Tjep. **Production |** Kloosterboer **Location |** Amsterdam, The Netherlands

VilaSofa

Tjep. was commissioned to create a shop environment for a new furniture brand called VilaSofa. VilaSofa is positioned between a conventional furniture shop and Ikea in terms of delivery speed. It offers mid-range prices for a broad customer base and a 48 hour delivery guarantee for all displayed models. The 48 hour claim became the main theme for the shop. Tjep. used the idea of a warehouse as a metaphor for speed. The conceptual solution was surprisingly simple and logical: a combination of warehouse esthetics and home esthetics. This resulted in the use of materials such as plywood, which is typically used for crates and combined with high-end glossy finishes. The symbols used in transportation and packaging have been metamorphosed into decorative elements that form room dividers, arranged to organize the space. A large wall suggests the idea of a magical villa. Is the customer looking at a giant styrofoam packaging element, or is the construction itself the actual villa? The designers created recesses for a big chandelier, a diversity of playfully arranged windows, romantic balconies, paintings, etc. They placed large picnic tables within the space for clients to sit down on and talk with the VilaSofa staff to become informed about products. They also designed mobile cashing systems so that clients are able to pay from their newly purchased sofa. VilaSofa offers a combination of existing sofa models and specially designed sofas by designers Monica Mulder and Khodi Feiz.

Design | SHH **Photography |** Gareth Gardner **Location |** Hampshire, UK

Applemore Technology College

The brief for the space was to provide a wide entrance to the dining area, with a sense of indoor greenery and some sheltered outdoor space, blurring the indoor/outdoor barriers. Dividing walls were to be translucent, increasing the sense of space and visibility; queue times were to decrease, there should be better provision for litter, and sound-absorbing materials were to be used.

The 4000sf interior has a relaxed cafeteria feel with zoned areas; some formal and some more casual. Overhead hanging graphic panels help to absorb noise. The Applemore College Canteen name has been shortened to ACC and used graphically throughout to create a strong identity. The space is double-height with the exposed ceiling workings lessened by being sprayed in dark grey, with a datum line all around at 2.5m height. Materials throughout are inspired by nature, with a yellow and green color theme, and have either natural patterns or natural colours.

An external shipping container provides cover for some extra external seating (sprayed green to add a visual link to the interior). Also outside, screened-off concrete stools (formed from Milton concrete tubes filled with concrete and aggregate) feature an iconic Belisha beacon at their centre.

Interior Design | Tobias Rehberger **Furniture Design |** Artek **Photography |** Wolfgang Günzel **Client |** La Biennale di Venezia/ 53rd International Art Exhibition **Location |** Venice, Italy

Cafeteria / Restaurant, Palazzo delle Esposizioni, Giardini della Biennale, Venice

Tobias Rehberger was awarded the Golden Lion for best artist at the 53rd International Art Exhibition, La Biennale di Venezia. Tobias Rehberger received the prize for the cafeteria, on which he collaborated closely with Artek using customized Artek furniture in an ingenious way. The cafeteria created by Rehberger is a complex scheme of geometric forms with contrasting colors. This visually disorienting environment draws from a specific example in the past: razzle dazzle or dazzle painting – a decorative style used on ships during the First World War.

TONIGHT

Index

Alessandro Suzzi, Italy

Alessandro Suzzi was born in Cosenza (Calabria) in 1980; shortly afterward, his family moved to Trani (Puglia).

He challenges his creativity with everything from comics, writing and oil painting to vector graphics for interior decoration.

His experience leads him to conceive of a wall (as well as a ceiling or a floor) as a huge piece of paper on which to design artistic embellishment for the house.

In 2007, he created the wall design group "Todesign," which specializes in wall design (a branch of interior design that involves the walls of an architectonic space).

Nowadays, he collaborates with several offices specialized in interior design to realize innovative showrooms through an esthetic process that evolves from the creation of the brand to the visual and chromatic conception of the walls.

P138-139, 140-141, 142-143, 144-145, 204-205

Ami Suma, Japan

Ami Suma was born in Tokyo, Japan in 1976 and later arrived in the USA. She has loved to draw pictures and decorate rooms since she was a child. In 2006, Ami started to make spontaneous "childhood imagination" driven art works. She was part of an exhibition called "Play things" at the Point of Contact gallery in Syracuse, New York and then featured in a group show called "Wonderland" at the Leo Fortuna gallery in Hudson, New York. She started designing murals for children's rooms in 2007. She wants to make the everyday life of children positive and imaginative.

P112-115

Artek, Finland

Artek was founded in 1935 by four young idealists: Alvar and Aino Aalto, Maire Gullichsen and Nils-Gustav Hahl. The business idea of the company was "to sell furniture and to promote modern culture of habitation by exhibitions and other means." The founders of Artek advocated a new kind of environment for everyday life. They believed in a grand synthesis of the arts and wanted to make a difference in town planning as well as architecture and design.

Today, Artek is renowned as one of the most innovative contributors to modern design, building on the heritage of Alvar Aalto. Functionality and timeless aesthetics are the essential elements in the creation of every product that bears the Artek name. Artek's comprehensive furniture system is based on the original idea of standards and systems, which has its origin from Alvar Aalto's L-system. The furniture range is versatile and can be customised for individual projects. Artek's furniture can be found in various types of spaces; from public spaces to homes, museums to schools and hotels to offices.

P228-233

b., Greece

Born in 1982, b. lives and works in Athens as an architect and street artist.

In a world full of corporate logos, religious symbols and team emblems, b. creates parasite brands, yellow and black mermaids, anchor-tattoos and oil-flags—personal symbols of his work that are seen on the walls of the city. He creates an art brand virus, which spreads in the urban landscape with colorful perseverance and a cheerful mood. He started as a graffiti writer in 1996 and studied architecture at the University of Volos from 2000 to 2006. He collaborates with "The Breeder" art gallery and has presented his work in numerous exhibitions, festivals and special commissions.

In 2007, his book "thisismybworld" was published by Futura/The Breeder and sponsored by Carhartt and Red Bull; it was curated by N. Argyropoulou and G.Tzirtzilakis, and included his street artwork in Athens and Berlin from 2002-2007.

P52-53, 54-57

Bitten Stetter, Germany

Bitten Stetter is a fashion designer from the University of Applied Science, Hamburg and has worked since 1999 as a self-employed fashion designer. In 2003, she established the international brand Bitten Stetter fashion and concept.

With 17 collections and more than 27 fashion shows, exhibitions and productions, Bitten Stetter moves between design and art. Her collections are present in international stores, yet are also on loan to museums. Various publications in relevant fashion and lifestyle magazines and international sales and production are the basis of her success.

At the centre of her work are a strong concept and staging. Her collections as well as her product line Fitting Tales thrive on stories, theme-worlds and small details, and go hand in hand with the concept store Fitting Forward.

Every piece and every garment plays a role in the creation of a special world. Nothing is accidental; a shirt is not only a shirt. Every bag and every button wants to tell a story for those who are willing to listen.

Parallel to her business, she has collaborated with brands such as adidas and MTV Germany. She is also a lecturer on fashion design at the AMD Hamburg, and for Style & Design as well as for the Master studies in Trends at the ZHDK in Zürich.

P166-167, 168-171, 172-173, 174-175, 176-177, 178-181

Branimir Sabljic, Croatia

Branimir Sabljic was born in Zagreb, Croatia, where he attended the high school for art and crafts. He presently is undertaking his final year of the Faculty of design, University of Zagreb - Department for Graphical Design. He worked for six years as a designer and art director in a design studio and an advertising agency before starting a career as a freelance artist and designer.

Branimir's work is largely influenced by graffiti. So far, he has created more than twenty large murals, gradually developing his "true 3D" style. He has also produced works in the fields of advertising, branding, exhibition design, interior design, painting, photo manipulation, and print and web design, and has organized several public graffiti-jam events for children and youth.

P18-21, 22-25

Camille Rousseau, France

Camille Rousseau grew up and attended school in central Paris and then went on to university in Arles. There she specialized in 3D animation at Supinfcom and subsequently worked for the Paris-based post production company VIEW as a motion designer, directing indents for French TV. In 2005, Camille began an MA program in Illustration at Central Saint Martins and graduated in June of 2007. Shortly afterwards, she was featured in Wallpaper*Magazine's 2009 Graduate Directory and went on to work for a wide range of international clients such as Dwell Magazine, EDF Energy, Diesel, Blueprint Magazine and ITV through LAMBIE NAIRN. She splits her time between London and Paris, and is based out of a studio in De Beauvoir Road, North London. She completed her first commission for Zizzi restaurant in 2009.

P80-81, 148-149, 150-153

David Lüpschen, Germany

David Lüpschen, also known as Jiar, was born in Cologne, Germany in 1979.

He is a graffiti artist, illustrator, and motion designer with his own label "Jiar" and is also the other half of the successful light writing crew Lichtfaktor.

Jiar started out by writing his name in letters. He then began to paint photo-realistically and soon progressed to developing his own characters.

His main concept is the combination and the contrast of clean photorealistic graffiti and dirty painted graphic elements and characters.

P32-33, 34-35, 36-39

Denise Fort, Germany

Denise Fort was born and raised in Munich, Germany, and spent time in Indonesia, India, Sri Lanka and Portugal. At the end of December 2008, she began traveling the world, performing odd jobs as an artist in private households and bars. In February 2009, she arrived in New Zealand and found the beautiful village Raglan. Enjoying the positive reaction and the open mind of New Zealanders towards her artwork, she has gained more confidence to spend time on working as an artist than ever before.

P70-73

Dizel&Sate, Sweden

Dizel&Sate is a Swedish design studio based in Stockholm. They have long been active within the world of alternative architecture, challenging traditional conceptions of minimalism and functionality. They have received commissions to design both retail and office spaces, as well as installations for public venues.

P162-163, 164-165, 186-189, 198-199, 120-123

Dopludo, Russia

Dopludo was founded by three friends in St. Petersburg, Russia in 2006. At first the team was named "two plus two is five." Later the name was shortened to "Dopludo," which is a phonetic transcription from French "deux plus deux," meaning "two plus two" in English. All of the collaborators work together on illustrations, murals, installations, and street projects in cold Russia and Scandinavia.

"In Dopludo we enjoy working with images from magic fairy tales, folk elements, mystical creatures and more or less abstract content... We like taking a position of looking at the world with the eyes of an unsophisticated child, who is still discovering new miracles. We care for everything genuine, pure & plain... Our main themes are integration and a fight against banality. We like working with people's reactions in the outside world as well as concentrating on complicated projects in the studio."

P60-65

Esther Stocker, Austria

Esther Stocker was born in Silandro, Italy and currently lives and works in Vienna.

She employs an optically complex spectrum of geometrized sign and grid systems in her artwork.

In 2004, she realized her first installation at AR/GE Kunst Galerie Museum, Bolzano.

Since then she has had numerous solo exhibitions at places including House of Art, České Budějovice, Galerie Krobath Wimmer, Vienna, CCNOA - Center for Contemporary Non-Objective Art, Brussels and Museum für Moderne Kunst Stiftung Ludwig, Vienna and group shows including "CELLA. Strukturen der Ausgrenzung und Disziplinierung," Complesso Monumentale di San Michele a Ripa Grande, Roma, "Good night, sleep tight," UpLoad Art Project, Trento, "Wiener Musterzimmer," Orangerie, Unteres Belvedere, Wien, "Beyond these Walls," South London Gallery, London, "Confines. Geography of the Boundary," IVAM Institut Valencià d'Art Modern, Valencia and "seeking constructive concrete structural 2," Vasarely Múzeum, Budapest.

P206-207, 208-209, 210-211, 212-213

Fabian Bertschinger, Switzerland

Fabian Bertschinger was born in Zürich, Switzerland in 1979.

He started drawing and painting letters as a kid and has always pursued this passion. He studied Graphic Design at the Lucerne University of Applied Sciences and Arts. Fabian is now doing postgraduate studies in Type Design and Typography (Master of Arts) at Zurich University of the Arts. He has worked as a graphic designer for 178 Aardige Ontwerpers in Amsterdam and for MetaDesign in Zürich. As a type-addicted artist, he has shaped letters for several independent art and design projects.

P66-69

Formavision, U.S.

Formavision is a New York-based experiential design studio that specializes in creating branded environments and cultural content to support and activate non-traditional marketing strategies. Since its inception in 2003, Formavision has been connecting its clients with key tastemakers and members of the creative community by conceiving and realizing projects that include the Diesel Denim Gallery, Lexus Light & Speed, the World of Coca-Cola, and the Starbucks Salon.

P98-99, 126-129, 182-185

Good Wives and Warriors, UK

Good Wives and Warriors is a creative partnership between Becky Bolton and Louise Chappell, who met while studying at the Glasgow School of Art. Rebecca means 'good wife' and Louise means 'warrior.' They have been working together on collaborative projects for the last 5 years and have recently been featured in Taschen's *Illustration Now Volume 3* and *IdN v16n2: Illustration Special* and have produced work for Adidas, Swatch and MTV. They have also exhibited wall paintings in Europe, South America, the USA and Australia. They are currently based in London.

P40-43

Hayon Studio, Spain

Spanish artist-designer Jaime Hayon was born in Madrid in 1974. As a teenager he submerged himself in skateboard culture and graffiti art, the foundation of the detailed, bold-yet-whimsical designs so distinctive in his work today. After studying industrial design in Madrid and Paris he joined Fabrica in 1997, working closely with legendary image maker and agitator Oliverio Toscani. In a short time he was promoted to the head of their design department, where he oversaw projects ranging from shops, restaurants, exhibition displays and designs to graphics. In 2004, Hayon Studio was founded.

P108-109, 110-111

Jutta Südbeck, Germany

Jutta Südbeck is a textile designer who works on the interface of colour, material, surface and space.

Her leitmotif led her from her studies in Hamburg (University of Applied Science), via New York (Aviva Stanoff Studio) and Westerstede (Kinnasand Interieur Textile GmbH) to Berlin, where she has worked freelance since 2005. She aims to bring together textile design, trend and concept.

Trend-counselor for entire collection concepts, she facilitates the design development of international companies, mainly in the field of product design with a focus on textiles. Recently, she has also been creating her own designs for fabrics and carpets, besides developing her unique product line. She is a lecturer on innovations in materials and textiles.

Jutta Südbeck stands for design with a vision, ideas with substance, and love of details.

P166-167, 168-171, 172-173, 174-175, 176-177, 178-181

Karim Rashid, U.S.

Karim Rashid is one of the most prolific designers of his generation. Over 3000 designs in production, over 300 awards and work in over 35 countries attest to Karim's legend of design. Karim has collaborated with clients to create democratic design for Method and Dirt Devil, furniture for Artemide and Magis, brand identity for Citibank and Hyundai, high tech products for LaCie and Samsung, and luxury goods for Veuve Clicquot and Swarovski, to name a few. Karim's work is featured in 20 permanent collections and he exhibits art in galleries worldwide. Karim is a perennial winner of the Red Dot award, Chicago Athenaeum Good Design award, I.D. Magazine Annual Design Review, and the IDSA Industrial Design Excellence award.

He holds honorary doctorates from the Ontario College of Art & Design and Corcoran College of Art & Design. Karim is a frequent guest lecturer at universities and conferences globally disseminating the importance of design in everyday life. Karim's work has been featured in magazines and books including "Time," "Financial Times," and countless more. In 2009, Rizzoli released Karim's latest monograph *KarimSpace*, featuring 36 of Karim's interior architecture designs. Other books include Karim's guide to living, *Design Your Self* (Harper Collins, 2006), *Digipop*, a digital exploration of computer graphics (Taschen, 2005), compact portfolio published by Chronicle Books (2004), as well as two monographs titled *Evolution* (Universe, 2004) and *I Want to Change the World* (Rizzoli, 2001). In his spare time Karim flirts with DJing, art and fashion and is determined to creatively touch every aspect of our physical landscape.

P118-121

Lovisa Burfitt, France

After graduating from the Beckmans School of Fashion in 1998, Lovisa Burfitt created the brand BURFITT in Sweden; it is an upper-range ready-to-wear line. In 2001, she decided to take a break and concentrate more on her work as a fashion illustrator. She was invited as a guest student to the Royal College of Art in Stockholm. After her studies she moved to Paris and in 2003 she re-launched the brand BURFITT in France, this time together with a business partner who oversees the commercial development of the brand. The concept of BURFITT is based on screen printed t-shirts and sweaters – created from original illustrations by Lovisa – and sold in limited editions. The woven part of the collection is more developed in terms of cut and choice of fabrics, and includes dresses, skirts, trousers and jackets. The brand incorporates both talents of Lovisa Burfitt: her illustrations and her creative ideas in terms of design.

Lovisa often collaborates with international stylists like Patti Wilson and Carine Roitfeld, who give her free reign to create masks and accessories for fashion stories to be published in "Vogue Paris," "Vogue Italia," "Numéro," "Harper's Bazaar."

Lovisa Burfitt continues to illustrate for various fashion magazines around the world, e.g. "Vogue," "ELLE," "Fashion Tale," "Cover," and "The End." She has also recently been collaborating with H&M "Graphic Communication for Flagship Stores," the Prêt-à-Porter fairs in Paris, Smart cars, and Bloomingdale's New York and she has been working on projects with Absolut Vodka and Dunkers Kulturhus for the exhibition "In Fashion – New Swedish Clothing Design." Her work has been displayed in many countries. From 2008 - 2009 her artwork was featured at the restaurant East in Stockholm in an exhibition titled 'Universe of Lovisa Burfitt – 100 m2 drawings'.

P74-75, 76-79

Mike Ballard, UK

Mike Ballard works across numerous visual media, including collage, painting, installation art and sculpture - along with film and music. He aims to challenge the viewer's perception of time and value.

Inspired by graffiti art and Hip-Hop music, both of which he has been producing since the mid-'80s, Ballard explores the connections between audio and visual forms. Sampling sounds and re-assembling discarded audio equipment into sculptures in antique form, he challenges the fragility of technology, creating bridges between old and new worlds.

Within his work, one medium becomes animated by another, creating appropriated hybrids. Found objects become electrified and integrate function and aesthetic value, each gaining a new status that questions the value of everything.

Assembling imagery from art history books and popular culture, Ballard's collages are then represented as illuminated signs, suggesting the possibility of a reconfigured time.

Ballard's paintings and prints challenge structures of perception, presenting powerful visions close to a hallucinogenic experience. This experience makes us question whether the world we see when we are "straight" is, in fact, the real world; or if it is somewhere outside, uncertain, mediated and not to be trusted.

P82-85

Miss Led, UK

An award-winning painter and illustrator both in London and Europe, Joanna Henly is barely in her 2nd year of working under the guise of Miss Led. In this time she has managed to amass a lucrative body of work with a high profile clientele. From painting live, a window for Selfridges and large-scale illustration work for Reebok and Diesel to painting bespoke interior pieces for celebrities and Brompton Rd boutiques, Led has created her distinctively playful, flirtatious and diverse styles on many surfaces in a plethora of mediums. Her illustrations have been commissioned; her artwork has been featured and published in magazines and newspapers nationwide. With a string of group shows under her belt she plans for a solo show early next year.

P46-51

Momorobo, Singapore

Momorobo is an art and design studio established in 2004, led and run by Morris Lee. It serves as a collaborative vehicle for like minded artists from various disciplines. The studio works on a variety of national, international and self initialized projects across various media like books, magazines, posters, typefaces, identities, music videos, exhibitions & mobile contents. Momorobo cohorts include Morris Lee, Eric Lee and Eddie Yeo.

P30-31

Mr. Important Design, U.S.

Mr. Important Design is an interior design firm that brings an ebullient perspective to interiors. They specialize in nightclubs, restaurants, bars and lounges, as well as select residential and retail projects.

P122-125

Pandarosa, Germany

Pandarosa, or the PANDAS as they are often referred to, is the brainchild created by Ariel Aguilera and Andrea Benyi. Originally hailing from Chile and Hungary, they joined forces in Melbourne and are currently based in Berlin.

Their work expands a wide range of approaches, including figurative illustration, abstract dreamscapes, bold iconic designs, detailed hand-drawn lines, nature inspired organic forms and pattern based imagery.

The duo's large-scale murals adorn the walls of Copenhagen's Hotel Fox and their 3D installation pieces have been featured at the Salon International de la Lingerie, Paris. They've decorated the pages of "Frame," "Icon," (inside) and Poster with their illustrations and have been utilised by various galleries, theatre and dance companies for their cultural graphic work. Clients such as Volkswagen, Domestic, Hecker Phelan Guthier, Lee Jeans, adidas and Julia DeVille have commissioned the pair & their work has been featured in titles such as "Grafik," "Tokion," "Artichoke" and "Design 360° " magazine, as well as many other book publications.

The group exhibits regularly and has been involved in numerous exhibitions across Melbourne, Hamburg, Madrid, Berlin, Barcelona, Tokyo and Los Angeles.

Pandarosa have a commitment to developing their own perspectives of thinking, feeling and researching their work and stimulating the imagination, thus showing the possibility of a productive coexistence of design and art.

P196-197

Pomme Chan, UK

Born and educated in Bangkok, Pomme relocated to London 7 years ago.

Pomme's first forays into professional freelancing led her straight into London's lifestyle magazine industry as an art director and illustrator. Over the past four years, her drawings have been featured in the Telegraph, "Grafik," and FT magazines, and she has worked on prestigious ad campaigns for the likes of Sony, Mercedes-Benz, VW, Microsoft, The Guardian, Nike and Topshop.

Using her unique felt-tip style, Pomme draws inspiration from nature, fashion, architecture, music and the female form.

Recently she has also begun exploring freehand typographical illustrations, several of which appeared in "La Perla Magazine," "New York Times Magazine," and others.

P146-147

Projekttriangle Design Studio, Germany

Projekttriangle Design Studio was established in 1999 by Prof. Danijela Djokic, Martin Grothmaak and Prof. Jürgen Späth.

Based in Stuttgart, Germany, the design studio works in applied and artistic projects from all disciplines of visual design: interactive design, information design and contemporary graphic design. In its applications and experiments, operating in the crossover zone between design, research and art, the studio executes international projects in the fields of graphic and corporate design, photography, interface design and media installations for industry, research and culture.

P9-13, 86-89, 90-93, 94-95

Qube Konstrukt, Australia

Founded in 2002, Qube Konstrukt is a multi-disciplinary design practice, whose creative diversity is the result of a shared vision to create a space in which inspiration and ideas are free to manifest themselves through different mediums.

While diversity defines their creative output, Qube Konstrukt's work is underpinned by a process-driven design philosophy that champions critical thought backed by research, experimentation and collaboration.

They work with confidence and ability across a broad range of disciplines encompassing identity and brand development, campaign, print and publication design, illustration and environmental graphics, as well as online, interactive and broadcast design. The studio is dedicated to developing unique, relevant and visually engaging projects.

P154-157

Rebecca Ward, U.S.

Rebecca Ward is an artist living and working in Austin, TX. She creates large scale site specific installations, primarily out of colored tapes. Most recently her work has been commissioned by Kate Spade, New York, where she produced installations in Kate Spade flagship stores in Tokyo, Japan and New York. Upcoming exhibitions include ROJO NOVA in Sao Paolo, Brazil (Summer 2010).

P216-217

ROW Studio, Mexico

ROW Studio is an architecture and design firm founded by Álvaro Hernández Felix and Alfonso Maldonado Ochoa in 2005. Its name derives from the term used in economy to refer to global phenomena (ROW: Rest of the World), as a critical standpoint against the clichés of Mexican architecture and the need to generate an international discourse both in processes and proposals departing from a multidisciplinary approach. At the same time, it addresses the need for expanding the field of study and influence from architecture to other disciplines such as sociology, psychology, economy, marketing, etc.

P190-193

SHH, UK

Architectural practice and design consultancy SHH was set up in 1992 by its three directors and co-founders: David Spence, Graham Harris and Neil Hogan, all of whom still actively run the company on a day-to-day basis, aided by six senior associates.

The company, based in its own west London studio, is a 40-strong practice and employs a highly talented and cosmopolitan team of architects, interior designers, graphic designers, and technical and administrative staff. Their work is also very much international in scope, located anywhere from Moscow to Bahrain, Pisa and Riga to Qatar, Lisbon and Tokyo. Their clients are prestige names anywhere in the world and include Christie's, Savills and the Mittal family.

P222-227

Spread Design, UK

Spread Design brings a new and inspired dimension to the design and delivery of 3D branded environments. Much of their work blurs the line between the traditional exhibition space and contemporary interior design.

Through specialist knowledge and experience that spans exhibitions and live events, to interiors and graphic design, they create dynamic world class environments that positively influence how people engage with and experience brands and organizations of all shapes and sizes.

P134-137

Steve Alexander (Rinzen), Germany

Steve Alexander is a member of Australian art and design collective Rinzen, and currently works from his studio in Berlin.

He has art-directed and illustrated campaigns, album covers and websites for Puma, Vodafone, Skittles, MTV, Orange, Warner Bros and "Wallpaper," and his signature fluid forms and typography have appeared on textiles, glasswear and a bicycle.

Steve has painted numerous large scale environments including galleries, a hotel, a bumper car arena, and the VW corporate headquarters in Wolfsburg.

His paintings have been exhibited as far afield as Tokyo, Madrid and Mexico City and he has conducted workshops and spoken at design conferences around the world.

P58-59

Studio Makkink & Bey, The Netherlands

Designer Jurgen Bey and architect Rianne Makkink have operated studio Makkink & Bey together in Rotterdam since 2002. The studio's many projects are very diverse, including public space projects, product design, architecture, exhibition design and applied arts. Aspects like analysing content, looking at the relationship between purpose and shape, design as story bearer, and the relationship between spectators and bearers, designer and commissioner, can originate from exhibition design and architecture, but are also used in projects of other disciplines.

P158-161

Takayuki Senzaki / TETUSIN DESIGN OFFICE, Japan

Born in 1978, Takayuki Senzaki had graduated from Kyushu Institute of Design. He established TETUSIN DESIGN OFFICE in Hakata, Fukuoka in 2009.

He does design direction for companies, stores, and the development of products.

Moreover, he creates wall graphics that are inspired by Japanese gardens and paintings of temples in various spaces.

P96-97, 100-101, 102-103, 106-107, 116-117

Tjep., The Netherlands

Tjep. works in a diversity of fields, from a Tiara for Princess Maxima to a nest consisting of giant rubber branches as a sofa, and from the development of campaign popcorn for Volkswagen to complete interiors and identities for restaurants, schools and shops. Tjep. was founded by Frank Tjepkema, a designer with an unpronounceable name that required a design intervention in itself, which resulted in Frank's agency name: Tjep.

Tjep. consists of a small team of highly motivated designers based in Amsterdam and working for companies such as Droog, British Airways, Camper, Heineken and Ikea.

P218-221

Trust in Design, France

Established in 2004 by Joran Briand, Etienne Vallet and Arthur de Chatelperron. Trust in Design is an interdisciplinary studio involving architecture, interior, graphic and product design. The studio is interested in the emerging relationships across different design disciplines and in the opportunities offered to designers by technological innovation. With collaborators specializing in different areas, each project is developed with an integral approach that challenges preconceptions and seeks alternative ways of living. Trust in Design is a collaboration that emerges from the awareness of the complexity of their world and attempts to make sense of it through the lens of design.

P104-105, 130-133

Via Grafik, Germany

Via Grafik is both a graphic design studio and an art collective based in Wiesbaden, Germany.

Working in the fields of graphic design, motion design, art direction and illustration, they are interested in developing design solutions at a unique visual level. Their aim hereby is to provide the best suitable design for the specific content of each individual project.

The realization of art projects increases Via Grafik's range of experience from which they can expand upon the creative process they use for commercial projects.

Their clients come from industries such as music, culture, art, and advertising, as well as youth culture oriented brands. They work for clients like Nike, adidas, Nintendo, MTV, VIVA, Junges Staatstheater Wiesbaden and We the People.

P44-45, 194-195

ZEK crew, Slovenia

The ZEK crew was established in 2001. It was primarily a graffiti crew that attempted street art and later shifted focus to design. Today, 9 members form this creative collective, including graphic, type, and industrial designers, graffiti writers, illustrators, an architect and a VJ. The crew has participated in several exhibitions and graffiti jams around Europe, with its work being featured in publications such as "Los Logos," "Stickeraward," "Graphotism" and "Concrete magazine." It is also famous for its StripGenerator, an award winning online comic generator. The crew always tries to incorporate wit, humor, and clever ideas into its artwork, mixing minimalism with kitsch, combining old styles with contemporary approaches, but not neglecting tradition through the use of improvisation. The variety of members enables the crew to take up any task and always remain in search of new visual sensations.

P14-17

ZMIK, Switzerland

ZMIK is a young studio for special design located in Basel, Switzerland. ZMIK explores opportunities at the fringes of various disciplines such as interior design, architecture, scenography and product design. Their work is focused on the design of communication, identity and individuality in space. ZMIK was founded in 2006 by Rolf Indermühle, Mattias Mohr and Magnus Zwyssig.

P26-29, 214-215

Acknowledgements

We would like to thank all the designers for their kind permission to publish their works, as well as all the photographers who have generously granted us the right to use their images. We are also very grateful to many other people whose names do not appear in the credits but who made specific contributions and provided support. Without these people, we would not be able to share these beautiful works of design with readers around the world.